V-Invex Functions and Vector Optimization

Optimization and Its Applications

VOLUME 14

Managing Editor
Panos M. Pardalos (University of Florida)

Editor—Combinatorial Optimization
Ding-Zhu Du (University of Texas at Dallas)

Advisory Board
J. Birge (University of Chicago)
C.A. Floudas (Princeton University)
F. Giannessi (University of Pisa)
H.D. Sherali (Virginia Polytechnic and State University)
T. Terlaky (McMaster University)
Y. Ye (Stanford University)

Aims and Scope
Optimization has been expanding in all directions at an astonishing rate during the last few decades. New algorithmic and theoretical techniques have been developed, the diffusion into other disciplines has proceeded at a rapid pace, and our knowledge of all aspects of the field has grown even more profound. At the same time, one of the most striking trends in optimization is the constantly increasing emphasis on the interdisciplinary nature of the field. Optimization has been a basic tool in all areas of applied mathematics, engineering, medicine, economics and other sciences.

The series *Optimization and Its Applications* publishes undergraduate and graduate textbooks, monographs and state-of-the-art expository works that focus on algorithms for solving optimization problems and also study applications involving such problems. Some of the topics covered include nonlinear optimization (convex and nonconvex), network flow problems, stochastic optimization, optimal control, discrete optimization, multi-objective programming, description of software packages, approximation techniques and heuristic approaches.

Shashi Kant Mishra, Shouyang Wang and
Kin Keung Lai

V-Invex Functions and Vector Optimization

 Springer

Shashi Kant Mishra
G.B. Pant Univ. of Agriculture & Technology
Pantnagar, India

Shouyang Wang
Chinese Academy of Sciences
Beijing, China

Kin Keung Lai
City University of Hong Kong
Hong Kong

Managing Editor:
Panos M. Pardalos
University of Florida

Editor/ Combinatorial Optimization
Ding-Zhu Du
University of Texas at Dallas

ISBN-13: 978-1-4419-4528-0   e-ISBN-13: 978-0-387-75446-8

Printed on acid-free paper.

9 8 7 6 5 4 3 2 1

springer.com

Preface

Generalizations of convex functions have previously been proposed by various authors, especially to establish the weakest conditions required for optimality results and duality theorems in nonlinear vector optimization. Indeed, these new classes of functions have been used in a variety of fields such as economics, management science, engineering, statistics and other applied sciences. In 1949 the Italian mathematician Bruno de Finetti introduced one of the fundamental generalized convex functions characterized by convex lower level sets, functions now known as "quasiconvex functions".

Since then other classes of generalized convex functions have been defined (not all useful at the same degree and with clear motivation) in accordance with the need of particular applications. In many cases such functions preserve some of the valuable properties of convex functions. One of the important generalization of convex functions is invex functions, a notion originally introduced for differentiable functions $f : C \rightarrow R$, C an open set of R^n, for which there exists some function $\eta : C \times C \rightarrow R^n$ such that $f(x) - f(y) \geq \eta(x, y)^T \nabla f(u)$, $\forall x, u \in C$. Such functions have the property that all stationary points are global minimizers and, since their introduction in 1981, have shown to be useful in a variety of applications. However, the major difficulty in invex programming problems is that it requires the same kernel function for the objective and constraints. This requirement turns out to be a severe restriction in applications. In order to avoid this restriction, Jeyakumar and Mond (1992) introduced a new class of invex functions by relaxing the definition invexity which preserves the sufficiency and duality results in the scalar case and avoids the major difficulty of verifying that the inequality holds for the same kernel function. Further, this relaxation allows one to treat certain nonlinear multiobjective fractional programming problems and some other classes of nonlinear (composite) problems. According to Jeyakumar and Mond (1992) A vector function $f : X \rightarrow R^p$ is said to be V-invex if there exist functions $\eta : X \times X \rightarrow R^n$ and $\alpha_i : X \times X \rightarrow R^+ - \{0\}$ such that for each

$x, \bar{x} \in X$ and for $i = 1, 2, ..., p$, $f_i(x) - f_i(\bar{x}) \geq \alpha_i(x, \bar{x}) \nabla f_i(\bar{x}) \eta(x, \bar{x})$.
For $p = 1$ and $\bar{\eta}(x, \bar{x}) = \alpha_i(x, \bar{x}) \eta(x, \bar{x})$ the above definition reduces to the usual definition of invexity given by Hanson (1981).

This book is concerned about the V-invex functions and its applications in nonlinear vector optimization problems. As we know that a great deal of optimization theory is concerned with problems involving infinite dimensional normed spaces. Two types of problems fit into this scheme are Variational and Control problems. As far as the authors are concerned this is the first book entirely concerned with V-invex functions and their applications.

Shashi Kant Mishra

Shouyang Wang

Kin Keung Lai

Contents

Chapter 1: General Introduction

1.1 Introduction

In many decision or design process, one attempts to make the best decision within a specified set of possible ones. In the sciences, "best" has traditionally referred to the decision that minimized or maximized a single objective optimization problem. But, we are rarely asked to make decisions based on only one objective, most often decisions are based on several usually conflicting objectives.

In nature, if the design of a system evolves to some final, optimal state, then it must include a balance for the interaction of the system with its surrounding-certainly a design based on a variety of objectives. Furthermore, the diversity of nature's design suggests infinity of such optimal states. In another sense, decisions simultaneously optimize a finite number of criteria, while there is usually infinity of optimal solutions. Multiobjective optimization provides the mathematical frame work to accommodate these demands.

The theory of multiobjective mathematical programming since it developed from multiobjective linear programming has been closely tied with convex analysis. Optimality conditions, duality theorems, saddle point analysis, constrained vector valued games and algorithms were established for the class of problems involving the optimization of convex objective functions over convex feasible regions. Such assumptions were very convenient because of the known separation theorems resulting from the Hahn-Banach theorem and the guarantee that necessary conditions for optimality were sufficient under convexity. However, not all practical problems, when formulated as multiobjective mathematical programs, fulfill the requirements of convexity, in particular, it was found that problems arising in economics and approximation theory could not be posed as convex programs. Fortunately, such problem were often found to have some characteristics in common with convex problems, and these properties could be exploited to establish theoretical results or develop algorithms. By abstraction, classes of functions having some useful properties shared with convexity could be defined. In fact, some notions of generalized convexity did exist before the need for it arose in mathematical programming, but it was through this need that researchers were given the incentive to develop a literature which has become extensive now, on the subject. At present there has been little unification of generalized convexity, although some notable exceptions are the papers of Schaible and Zimba (1981) the wide ranging work of Avriel, Diwert, Schaible and Zang (1988) and Jeyakumar and Mond (1992).

1.2 Multiobjective Programming Problems

The general multiobjective programming model can be written as

(VP) $V - \text{Minimize} \quad \left(f_1(x), ..., f_p(x)\right)$

subject to $g(x) \le 0$,

where $f_i : X_0 \rightarrow R$, $i = 1, ..., p$ and $g : X_0 \rightarrow R^m$ are differentiable functions on $X_0 \subseteq R^n$ open. Note here that the symbol "$V - \text{Minimize}$" stands for vector minimization. This is the problem of finding the set of weak minimum/efficient/properly efficient/conditionally properly efficient (Section 4 of the present Chapter) points for (VP). When $p = 1$, the problem (VP) reduces to a scalar optimization problem and it is denoted by (P). Convexity of the scalar problem (P) is characterized by the inequalities:

$$f(x) - f(u) - f'(u)(x - u) \ge 0$$
$$g(x) - g(u) - g'(u)(x - u) \ge 0,$$
$$\forall \, x, u \in X_0.$$

Hanson (1981) observed that the functional form $(x - u)$ here plays no role in establishing the following two well-known properties in scalar convex programming:

(S) Every feasible Kuhn-Tucker point is global minimum.

(W) Weak duality holds between (P) and its associated dual problem.

Having this in mind, Hanson (1981) considered problem (P) for which there exists a function $\eta : X_0 \times X_0 \rightarrow R^n$ such that

(I) $f(x) - f(u) - f'(u)\eta(x, u) \ge 0$

$g(x) - g(u) - g'(u)\eta(x, u) \ge 0, \quad \forall \, x, u \in X_0,$

and showed that such problems (known now as invex problems [Craven (1981, 1988)]) also possess properties (S) and (W). Since then, various generalizations of conditions (I) to multiobjective problems and many properties of functions that satisfy (I) have been established in the literature, e.g. Ben-Israel and Mond (1986), Craven (1988), Craven and Glover (1985), Martin (1985). However, the major difficulty is that the invex problems require the same kernel function $\eta(x, u)$ for the objective and the constraints. This requirement turns out to be a severe restriction in applications. Because of this restriction, pseudo-linear multiobjective problems (Chew and Choo (1984)) and certain nonlinear multiobjective fractional programming problems require separate treatment as far as optimality and duality properties are concerned. In order to avoid this restriction, Jeyakumar and Mond (1992) introduced a new class of functions, which we shall present in the next Section. We have developed necessary and sufficient optimality conditions of the minimization problem involving differentiable and non-differentiable functions in the subsequent chapters. We have discussed nonsmooth problems and compared with minimax problems and further we have presented nonsmooth composite problems and discussed optimality, duality and saddle point analysis. We have also considered multiobjective continuous and control problems in Chapter six and established sufficient optimality conditions and duality results.

1.3 *V* – Invexity

Jeyakumar and Mond (1992) introduced the notion of V-invexity for a vector function $f = (f_1, f_2, ..., f_p)$ and discussed its applications to a class of constrained multiobjective optimization problems. We now give the definitions of Jeyakumar and Mond (1992) as follows.

Definition 1.3.1: A vector function $f : X \to R^p$ is said to be *V-invex* if there exist functions $\eta : X \times X \to R^n$ and $\alpha_i : X \times X \to R^+ - \{0\}$ such that for each $x, \bar{x} \in X$ and for $i = 1, 2, ..., p$,
$$f_i(x) - f_i(\bar{x}) \geq \alpha_i(x, \bar{x}) \nabla f_i(\bar{x}) \eta(x, \bar{x}).$$
For $p = 1$ and $\bar{\eta}(x, \bar{x}) = \alpha_i(x, \bar{x}) \eta(x, \bar{x})$ the above definition reduces to the usual definition of invexity given by Hanson (1981).

Definition 1.3.2: A vector function $f : X \to R^p$ is said to be *V-pseudoinvex* if there exist functions $\eta : X \times X \to R^n$ and
$$\beta_i : X \times X \to R^+ - \{0\}$$

such that for each $x, \bar{x} \in X$ and for $i = 1, 2, ..., p$,

$$\sum_{i=1}^{p} \nabla f_i(\bar{x}) \eta(x, \bar{x}) \geq 0 \Rightarrow \sum_{i=1}^{p} \beta_i(x, \bar{x}) f_i(x) \geq \sum_{i=1}^{p} \beta_i(x, \bar{x}) f_i(\bar{x}).$$

Definition 1.3.3: A vector function $f : X \to R^p$ is said to be *V-quasiinvex* if there exist functions $\eta : X \times X \to R^n$ and $\delta_i : X \times X \to R^+ - \{0\}$ such that for each $x, \bar{x} \in X$ and for $i = 1, 2, ..., p$,

$$\sum_{i=1}^{p} \delta_i(x, \bar{x}) f_i(x) \leq \sum_{i=1}^{p} \delta_i(x, \bar{x}) f_i(\bar{x}) \Rightarrow \sum_{i=1}^{p} \nabla f_i(\bar{x}) \eta(x, \bar{x}) \leq 0.$$

It is evident that every V-invex function is both V-pseudo-invex (with $\beta_i(x, \bar{x}) = \dfrac{1}{\alpha_i(x, \bar{x})}$) and V-quasiinvex (with $\delta_i(x, \bar{x}) = \dfrac{1}{\alpha_i(x, \bar{x})}$). Also if we set

$$p = 1, \alpha_i(x, \bar{x}) = 1, \ \beta_i(x, \bar{x}) = 1, \delta_i(x, \bar{x}) = 1 \text{ and } \eta(x, \bar{x}) = x - \bar{x},$$

then the above definitions reduce to those of convexity, pseudo-convexity and quasi-convexity, respectively.

Definition 1.3.4: A vector optimization problem:
(VP) $V - \min(f_1, f_2, ..., f_p)$

subject to $g(x) \leq 0$,

where $f_i : X \to R, \ i = 1, 2, ..., p$ and $g : X \to R^m$ are   differentiable functions on X, is said to be *V-invex vector optimization problem* if each $f_1, f_2, ..., f_p$ and $g_1, g_2, ..., g_m$ is a V-invex function.

Note that, invex vector optimization problems are necessarily V-invex, but not conversely. As a simple example, we consider following example from Jeyakumar and Mond (1992).

Example 1.3.1: Consider

$$\min_{x_1, x_2 \in R} \left(\frac{x_1^2}{x_2}, \frac{x_1}{x_2} \right)$$

subject to $1 - x_1 \leq 1,$
$1 - x_2 \leq 1.$

Then it is easy to see that this problem is a V-invex vector optimization problem with $\alpha_1 = \dfrac{\bar{x}_2}{x_2}, \alpha_2 = \dfrac{\bar{x}_1}{x_1}, \beta_1 = 1 = \beta_2$ and $\eta(x,\bar{x}) = x - \bar{x}$; but clearly, the problem does not satisfy the invexity conditions with the same η.

It is also worth noticing that the functions involved in the above problem are invex, but the problem is not necessarily invex.

It is known (see Craven (1981)) that invex problems can be constructed from convex problems by certain nonlinear coordinate transformations. In the following, we see that V-invex functions can be formed from certain nonconvex functions (in particular from convex-concave or linear fractional functions) by coordinate transformations.

Example 1.3.2: Consider function, $h : R^n \rightarrow R^p$ defined by
$$h(x) = \left(f_1(\phi(x)),..., f_p(\phi(x)) \right),$$
where $f_i : R^n \rightarrow R$, $i = 1,2,..., p$, are strongly pseudo-convex functions with real positive functions $\alpha_i, \phi : R^n \rightarrow R^n$ is surjective with $\phi'(\bar{x})$ onto for each $\bar{x} \in R^n$. Then, the function h is V-invex.

Example 1.3.3: Consider the composite vector function
$$h(x) = \left(f_1(F_1(x)),..., f_p(F_p(x)) \right),$$
where for each $i = 1, 2,..., p$, $F_i : X_0 \rightarrow R$ is continuously differentiable and pseudolinear with the positive proportional function $\alpha_i(\cdot, \cdot)$, and $f_i : R \rightarrow R$ is convex. Then, $h(x)$ is $V - $ invex with $\eta(x,y) = x - y$. This follows from the following convex inequality and pseudolinearity conditions:
$$f_i(F_i(x)) - f_i(F_i(y)) \geq f_i'(F_i(y))(F_i(x) - F_i(y))$$
$$= f_i'(F_i(y))\alpha_i(x,y)F_i'(y)(x - y)$$
$$= \alpha_i(x,y)(f_i \circ F_i)'(y)(x - y).$$
For a simple example of a composite vector function, we consider
$$h(x_1, x_2) = \left[e^{x_1/x_2}, \frac{x_1 - x_2}{x_1 + x_2} \right],$$
where $x_0 = \left\{ (x_1, x_2) \in R^2 : x_1 \geq 1, x_2 \geq 1 \right\}.$

Example 1.3.4: Consider the function
$$H(x) = \left(f_1\left((g_1 \circ \psi)(x)\right), \ldots, f_p\left((g_p \circ \psi)(x)\right)\right),$$
where each f_i is pseudolinear on R^n with proportional functions $\alpha_i(x, y)$, ψ is a differentiable mapping from R^n onto R^n such that $\psi'(y)$ is surjective for each $y \in R^n$, and $f_i : R \to R$ is convex for each i. Then H is V – invex.

Jeyakumar and Mond (1992) have shown that the V – invexity is preserved under a smooth convex transformation.

Proposition 1.3.1: Let $\psi : R \to R$ be differentiable and convex with positive derivative everywhere; let $h : X_0 \to R^p$ be V – invex. Then, the function $h_\psi(x) = \left(\psi(h_1(x)), \ldots, \psi(h_p(x))\right)$, $x \in X_0$ is V – invex.

Proof: Let $x, u \in X_0$. Then, from the monotonicity of ψ and V – invexity of h, we get
$$\psi(h_i(x)) \geq \psi\left(h_i(u) + \alpha_i(x, u)h_i'(u)\eta(x, u)\right)$$
$$\geq \psi(h_i(u)) + \psi'(h_i(u))\alpha_i(x, u)h_i'(u)\eta(x, u)$$
$$= \psi(h_i(u)) + \alpha_i(x, u)(\psi \circ h_i)'(u)\eta(x, u).$$
Thus, $h_\psi(x)$ is V – invex.

Recall that a point $u \in R^n$ is said to be a (global) *weak minimum* of a vector function $f : R^n \to R^p$ if there exists no $x \in R^n$ for which
$$f_i(x) < f_i(u), \quad i = 1, \ldots, p.$$
The following very important property of V – invex functions was also established by Jeyakumar and Mond (1992).

Proposition 1.3.2: Let $f : R^n \to R^p$ be V – invex. Then $y \in R^n$ is a (global) weak minimum of f if and only if there exists
$$0 \neq \tau \in R^p, \tau \geq 0, \sum_{i=1}^{p} \tau_i f_i'(y) = 0.$$

Proof: $(\Rightarrow)$ Suppose that u is weak minimum for f. Then the following linear system $x \in R^n$, $f_i(u) < f_i(x)$, $i = 1, \ldots, p$, is inconsistenct. Hence, the conclusion follows from the Gordan Alternative Theorem(Craven (1978)).

($\Leftarrow$) Assume that $\sum_{i=1}^{p} \tau_i f_i'(y) = 0$, for some $0 \neq \tau \in R^p, \tau \geq 0$. Suppose that the point u is not a weak minimum for f. Then there exists $x_0 \in R^n$ such that $f_i(x_0) < f_i(u)$, $i = 1, ..., p$. Since f is $V - $ invex, there exists $\alpha_i(x_0, u) > 0$, $i = 1, ..., p$ and $\eta(x_0, u) \in R^n$ such that

$$\frac{1}{\alpha_i(x_0, u)}(f_i(x_0) - f_i(u)) \geq f_i'(u)\eta(x_0, u).$$

So,

$$\sum_{i=1}^{p} \frac{1}{\alpha_i(x_0, u)}(f_i(x_0) - f_i(u)) < 0,$$

and hence

$$\sum_{i=1}^{p} \tau_i f_i'(u)\eta(x_0, u) < 0.$$

This is a contradiction.

By Proposition 1.3.2, one can conclude that for a $V - $ invex vector function every critical point $\left(\text{i.e. } f_i'(y) = 0, i = 1, ..., p\right)$ is a global weak minimum.

Hanson *et al.* (2001) extended the (scalarized) generalized type-I invexity into a vector (V-type-I) invexity.

Definition 1.3.5: The vector problem (VP) is said to be *V-type-I* at $\bar{x} \in X$ if there exist positive real-valued functions α_i and β_j defined on $X \times X$ and an $n - $ dimensional vector-valued function $\eta : X \times X \rightarrow R^n$ such that

$$f_i(x) - f_i(\bar{x}) \geq \alpha_i(x, \bar{x})\nabla f_i(\bar{x})\eta(x, \bar{x})$$

and

$$-g_j(\bar{x}) \geq \beta_j(x, \bar{x})\nabla g_j(\bar{x})\eta(x, \bar{x}),$$

for every $x \in X$ and for all $i = 1, 2, ..., p$ and $j = 1, 2, ..., m$.

Definition 1.3.6: The vector problem (VP) is said to be *quasi-V-type-I* at $\bar{x} \in X$ if there exist positive real-valued functions α_i and β_j defined on $X \times X$ and an $n - $ dimensional vector-valued function $\eta : X \times X \rightarrow R^n$ such that

$$\sum_{i=1}^{p} \tau_i \alpha_i(x, \bar{x})[f_i(x) - f_i(\bar{x})] \leq 0 \Rightarrow \sum_{i=1}^{p} \tau_i \eta(x, \bar{x})\nabla f_i(\bar{x}) \leq 0$$

and

$$-\sum_{j=1}^{m}\lambda_j\beta_j(x,\bar{x})g_j(\bar{x}) \leq 0 \Rightarrow \sum_{j=1}^{m}\lambda_j\eta(x,\bar{x})\nabla g_j(\bar{x}) \leq 0,$$

for every $x \in X$.

Definition 1.3.7: The vector problem (VP) is said to be *pseudo-V-type-I* at $\bar{x} \in X$ if there exist positive real-valued functions α_i and β_j defined on $X \times X$ and an $n-$ dimensional vector-valued function $\eta : X \times X \rightarrow R^n$ such that

$$\sum_{i=1}^{p}\tau_i\eta(x,\bar{x})\nabla f_i(\bar{x}) \geq 0 \Rightarrow \sum_{i=1}^{p}\tau_i\alpha_i(x,\bar{x})[f_i(x) - f_i(\bar{x})] \geq 0$$

and

$$\sum_{j=1}^{m}\lambda_j\eta(x,\bar{x})\nabla g_j(\bar{x}) \geq 0 \Rightarrow -\sum_{j=1}^{m}\lambda_j\beta_j(x,\bar{x})g_j(\bar{x}) \geq 0,$$

for every $x \in X$.

Definition 1.3.8: The vector problem (VP) is said to be *quasi-pseudo-V-type-I* at $\bar{x} \in X$ if there exist positive real-valued functions α_i and β_j defined on $X \times X$ and an $n-$ dimensional vector-valued function $\eta : X \times X \rightarrow R^n$ such that

$$\sum_{i=1}^{p}\tau_i\alpha_i(x,\bar{x})[f_i(x) - f_i(\bar{x})] \leq 0 \Rightarrow \sum_{i=1}^{p}\tau_i\eta(x,\bar{x})\nabla f_i(\bar{x}) \leq 0$$

and

$$\sum_{j=1}^{m}\lambda_j\eta(x,\bar{x})\nabla g_j(\bar{x}) \geq 0 \Rightarrow -\sum_{j=1}^{m}\lambda_j\beta_j(x,\bar{x})g_j(\bar{x}) \geq 0,$$

for every $x \in X$.

Definition 1.3.9: The vector problem (VP) is said to be *pseudo-quasi-V-type-I* at $\bar{x} \in X$ if there exist positive real-valued functions α_i and β_j defined on $X \times X$ and an $n-$ dimensional vector-valued function $\eta : X \times X \rightarrow R^n$ such that

$$\sum_{i=1}^{p}\tau_i\eta(x,\bar{x})\nabla f_i(\bar{x}) \geq 0 \Rightarrow \sum_{i=1}^{p}\tau_i\alpha_i(x,\bar{x})[f_i(x) - f_i(\bar{x})] \geq 0$$

and

$$-\sum_{j=1}^{m}\lambda_j\beta_j(x,\bar{x})g_j(\bar{x})\le 0 \Rightarrow \sum_{j=1}^{m}\lambda_j\eta(x,\bar{x})\nabla g_j(\bar{x})\le 0,$$

for every $x \in X$.

Nevertheless the study of generalized convexity of a vector function is not yet sufficiently explored and some classes of generalized convexity have been introduced recently. Several attempts have been made by many authors to introduce possibly a most wide class of generalized convex function, which can meet the demand of a real life situation to formulate a nonlinear programming problem and therefore get a best possible solution for the same.

1.4 Efficient Solution for Optimal Problems with Multicriteria

For the vector function $f(x)=(f_1(x),...,f_p(x))$ and a set of feasible point $K \subseteq R^n$ for which it is desirable to minimize$f(x)$, (maximize$f(x)$). x^0 is defined to be *efficient* if $x^0 \in K$ and there is no other $x \in K$ such that $f(x)\le f(x^0)(f(x)\ge f(x^0))$.

The properness of the efficient solution of the optimal problem with multicriteria has been introduced at the early stage of the study of this problem (Kuhn and Tucker (1951)). Geoffrion (1968) defined the properness for the purpose of eliminating an undesirable possibility in the concept of efficiency, namely the possibility of the criterion functions being such that efficient solutions could be found for which the marginal gain for one function could be made arbitrarily large relative to the marginal losses for the others. Geoffrion (1968) gave a theorem describing the relation of the Kuhn-Tucker proper efficient solutions and his proper efficient solution.

In this section, we summarize briefly the known results of proper (improper) efficient solutions for (VP), and apply them to five examples.

The problems discussed in the papers of Kuhn-Tucker (1951), Geoffrion (1968), Tamura and Arai (1982) and Singh and Hanson (1991) are of the following nature:

$$V - \text{Maximize } f(x)$$
$$\text{subject to } g(x)\le 0,$$

where $f(x)$ and $f(x)\le f(x^0)$ are $g(x)$ are $p-$ dimensional and $m-$ dimensional vectors.

Let K denote the set of feasible solutions of the above vector maximum problem.

Definition 1.4.1 [Kuhn-Tucker (1951)]:
An efficient solution x^0 is called a proper efficient solution if there exists no $x \in K$ such that

$$\nabla f(x^0)x \geq 0, \ \nabla g_I(x^0)x \geq 0,$$

where $\nabla g_I(x^0)$ is a matrix whose row vector is a gradient function of an active constraint. We call this solution a *KT-proper efficient solution.*

Definition 1.4.2 [Geoffrion (1968)]: An efficient solution x^0 is called a proper efficient solution if there exists a scalar $M > 0$ such that, for each i,

$$\frac{f_i(x) - f_i(x^0)}{f_j(x^0) - f_j(x)} \leq M$$

for some j such that $f_j(x) < f_j(x^0)$, whenever, $x \in K$ and

$$f_i(x) > f_i(x^0)$$

For minimization problem (VP) we have the following inequality for $M > 0$

$$\frac{f_i(x^0) - f_i(x)}{f_j(x) - f_j(x^0)} \leq M$$

for some j such that $f_j(x) > f_j(x^0)$, whenever x is feasible for (VP) and $f_i(x) < f_i(x^0)$.

We call this solution a G-proper efficient solution.

Proposition 1.4.1 [Geoffrion (1968)]: Assume that the Kuhn-Tucker constraint qualification holds at x^0. Then a G-proper efficient solution implies a KT-proper efficient solution.

Now, let is examine the proper and improper efficient solutions of the following example in some detail.

Example 1.4.1 [Kuhn-Tucker (1951)] The problem considered is as follows:

$$\text{Maximize}(x, -x^2 + 2x)$$
$$\text{subject to} \ \ 2 - x \geq 0, \ \ x \geq 0.$$

The feasible region and the functions are shown in the fig 1a:

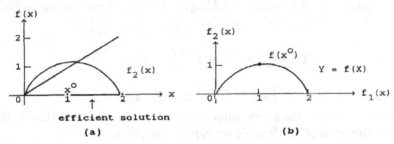

Fig. 1. (a) Functions $f_1(x)$, $f_2(x)$, and set of efficient solutions. (b) $Y = f(X)$ in example 1.4.1.

In Geoffrion's (1968) definition of proper efficiency M is independent of x, and it may happen that if f is unbounded such an M may not exist. Also an optimizer might be willing to trade different levels of losses for different levels of gains by different values of the decision variable x. Singh and Hanson (1991) extended the concept to situation where M depends on x.

Definition 1.4.3 [Singh and Hanson (1991)]: The point x^0 is said to be conditionally properly efficient for (VP) if x^0 is efficient for (VP) and there exists a positive function $M(x)$ such that, for each i, we have

$$\frac{f_i(x) - f_i(x^0)}{f_j(x^0) - f_j(x)} \le M(x),$$

for some j such that $f_j(x) < f_j(x^0)$ whenever $x \in X$ and

$$f_i(x) > f_i(x^0).$$

For $V-\text{Min}$ problem the above definition can be stated as: The point x^0 is said to be conditionally properly efficient for $(V-\text{Min})$ if x^0 is efficient for $(V-\text{Min})$ and there exists a positive function $M(x)$ such that, for i, we have

$$\frac{f_i(x^0) - f_i(x)}{f_j(x) - f_j(x^0)} \le M(x),$$

for some j such that $f_j(x) > f_j(x^0)$ whenever $x \in X$ and

$$f_i(x) < f_i(x^0).$$

The following example is from Mishra and Mukherjee (1995a).

Example 1.4.4. [Mishra and Mukherjee (1995a)]: Consider the problem
$$V - \text{Minimize} \left(f_1(x_1, x_2), f_2(x_1, x_2) \right)$$

where $f_1(x_1, x_2) = \dfrac{x_1}{x_2}$ and $f_2(x_1, x_2) = \dfrac{x_2}{x_1}$

subject to $(x_1, x_2) \in R^2, 1 - x_1 \le 0, 1 - x_2 \le 0.$

It can be shown that every point of the feasible region is efficient. But, none of the point of the feasible set is properly efficient.

Let $x^* = (a, b)$ be an efficient solution. By symmetry of the program, we may assume that $1 \le a \le b$. Let M be any positive number. Choose $x = (x_1, x_2)$ so that

$$\frac{x_2}{x_1} > Max\left(M, \frac{b}{a} \right).$$

Then

$$f_2(x) = \frac{x_2}{x_1} > \frac{b}{a} = f_2(x^*),$$

but

$$\frac{f_2(x^*) - f_2(x)}{f_1(x) - f_1(x^*)} = \frac{\dfrac{b}{a} - \dfrac{x_2}{x_1}}{\dfrac{x_1}{x_2} - \dfrac{a}{b}} = \frac{bx_2(bx_1 - ax_2)}{ax_1(bx_1 - ax_2)} = \frac{bx_2}{ax_1} \ge \frac{x_2}{x_1} > M.$$

This shows that $x^* = (a, b)$ can not be properly efficient. But, every efficient solution is conditionally properly efficient:

Choose $M(x) \ge \dfrac{bx_2}{ax_1}$, where $x = (x_1, x_2)$, then

$$\frac{f_2(x^*) - f_2(x)}{f_1(x) - f_1(x^*)} = \frac{bx_2}{ax_1} \le M(x),$$

$f_2(x) = \dfrac{x_2}{x_1} > \dfrac{b}{a} = f_2(x^*)$, where (x_1, x_2) is feasible and

$$f_1(x) = \frac{x_1}{x_2} < \frac{a}{b} = f_1(x^*).$$

Thus, x^* is conditionally properly efficient.

Chapter 2: V-Invexity in Nonlinear Multiobjective Programming

2.1 Introduction

Hanson's (1981) introduction of invex functions was motivated by the question of finding the widest class of functions for which weak duality hold for dual programs, such as the Wolfe and Mond-Weir duals, formulated from the necessary optimality conditions. Since then, various generalizations of invexity have been introduced in the literature e.g. Craven and Glover (1985), Egudo (1989), Hanson and Mond (1982), Kaul and Kaur (1985), Martin (1983), Kaul, Suneja and Srivastava (1994), Jeyakumar and Mond (1992), Mond and Hanson (1984, 1989), Nanda and Das (1993, 1994), Smart (1990), Weir (1988), Rueda and Hanson (1988), Mond and Husain (1989), Mond, Chandra and Husain (1988), Mishra and Mukherjee (1994a, 1994b, 1995, 1996, 1996a, 1996b).

However, the major difficulty is that the invex problems require the same kernel function for the objective and the constraints. This requirement turns out to be a severe restriction in applications. Because of this restriction, pseudolinear multiobjective problems (Chew and Choo (1984), Rueda (1989), Kaul, Suneja and Lalitha (1993), Komlosi (1993), Mishra (1995c), Mishra and Mukherjee (1996b) and certain nonlinear multiobjective fractional programming problems require separate treatment as far as optimality and duality properties are concerned.

In this Chapter, we consider the role of invexity and its generalizations, namely V-pseudo-invexity and V-quasi-invexity in standard multiobjective programming; in particular, the replacement is made of invexity in results related to necessary and sufficient optimality conditions, duality theorems symmetric duality results and vector valued constrained games. A vast number of theorems developed during the evolution of nonlinear programming theory were stated with assumptions of invexity. In most cases it has been possible to generalize these results under the assumptions of V-invexity. However, this has not been a direct process. Intermediate and overlapping results have been achieved using the various notions of generalized convexity discussed in Chapter 1.

Following Mangasarian's (1969) use of pseudo-convexity and quasi-convexity for optimality and duality theorems, duality theory has been constructed based on particular generalizations of convexity; see, for example, the works of Crouzeix (1981) on quasi-convex functions, Avriel (1979) on (h, F)-convex functions, Preda (1992) on (F, p)-convex, Preda (1994) on (F, p)-quasi-convex Mangasarian, (F, p)-quasi-convex Ponstein.

There has been substaintial progress made by authors such as Craven and Glover (1981), Mond and Hanson (1984), Martin (1985) in developing a complete duality theory using invex functions, and by Jeyakumar and Mond (1992), and Mishra (1995a) using V-invex functions. Here, we outline the relationship of V-invexity to Mond-Weir duals, via Kuhn-Tucker conditions. Next, the necessary and sufficient optimality conditions for a class of nondifferentiable multiobjective programming problem will be established. In section six, vector valued infinite game is associated to a pair of multiobjective programming problem and finally in the last section a multiobjective symmetric duality theorem is established.

The general nonlinear multiobjective program to be considered is:

(VP): $\qquad V - Min\left(f_1(x), \ldots, f_p(x) \right)$

$$\text{subject to} \quad g(x) \le 0,$$

where $f_i : X_0 \to R$, $i = 1, \ldots, p$ and $g : X_0 \to R^m$ are differentiable functions on $X_0 \subseteq R^n$ open. When $p = 1$, the problem (VP) reduces to a single objective case and gives (P) of Wolfe (1961), Avriel (1976), Kaul and Kaur (1985).

It is assumed that the program (VP) contains no equality constraints; equality constraints of the form $h(x) = 0$ could be re-written as $h(x) \ge 0$, $-h(x) \ge 0$ in order to put an equality constrained optimization problem in form of (VP).

The Fritz-John type necessary conditions for a feasible point x^* to be optimal for (VP) are (John (1948)) the existence of $\tau \in R^p$, $\lambda \in R^m$ such that

$$\sum_{i=1}^{p} \tau_i f_i'\left(x^*\right) + \sum_{j=1}^{m} \lambda_j g_j'\left(x^*\right) = 0 \tag{2.1}$$

$$\lambda_j g_j\left(x^*\right) = 0, \quad j = 1, \ldots, m, \tag{2.2}$$

$$\tau \ge 0, \quad \lambda \ge 0. \tag{2.3}$$

There are no restrictions on the objective or constraint functions apart from differentiability.

However, by imposing a regularity condition on the constraint functions, the $\tau \in R^p$ may without loss of generality, be taken as $\sum_{i=1}^{p} \tau_i = 1$, and we obtain the Kuhn-Tucker type conditions (Kuhn-Tucker (1951)) or the weak Arrow-Hurwicz-Uzawa constraint qualification (Mangasarian (1969)):

There exist $\tau \in R^p$, $\lambda \in R^m$ such that

$$\sum_{i=1}^{p} \tau_i f_i'\left(x^*\right) + \sum_{j=1}^{m} \lambda_j g_j'\left(x^*\right) = 0 \qquad (2.4)$$

$$\lambda_j g_j\left(x^*\right) = 0, \quad j = 1,...,m, \qquad (2.5)$$

$$\tau \geq 0, \quad \sum_{i=1}^{p} \tau_i = 1, \quad \lambda \geq 0. \qquad (2.6)$$

It is shown in Mangasarian (1969) that the Kuhn-Tucker type conditions are necessary for optimality regardless of any convexity conditions on g.

2.2 Sufficiency of the Kuhn-Tucker Conditions

Kuhn and Tucker (1951) prove that when f is differentiable and convex, and g is differentiable and concave, then a feasible point x^* of (P), for which there exists some $\tau^* \in R^m$ such that $\left(x^*, \tau^*\right)$ satisfy the Kuhn-Tucker conditions, is an optimal solution of (P).

Mangasarian (1969) weakened the convexity requirements for this result to hold; it is sufficient that f be pseudo-convex and $g_j, j \in J$, $J = \left\{ j : g_j(x^*) = 0 \right\}$ be differentiable and quasi-concave.

The question of which is the widest class of functions giving sufficiency of the Kuhn-Tucker conditions is used to introduce $V - $ invex functions in Jeyakumar and Mond (1992). There, it is shown that sufficiency follows when $\tau_i f_i$ is $V - $ pseudo-invex, $i = 1,...,p$, and $\lambda_j g_j$ is $V - $ quasi-invex, $j = 1,...,m$, with respect to the same η.

The concept of efficiency or Pareto optimality in multiobjective programming has important role in all optimal decision problems with noncomparable criteria. Geoffrion (1968) introduced a slightly restricted definition of efficiency called proper efficiency for the purpose of eliminating efficient points of a certain anomalous type that lends itself to more satisfactory characterization. Many researchers have obtained necessary and sufficient conditions of Kuhn-Tucker type for a feasible point to be properly efficient, for example see Kaul, Suneja and Srivastava (1994) and references therein. Singh and Hanson (1991) pointed out that M involved in the definition of proper efficiency (Chapter 1, Section 4) is independent of x and it may happen that if f is unbounded such an M may not exist. Hence they generalized the definition to cover situations where Geoffrion's (1968) definition does not apply.

In light of above discussion we establish the following Kuhn-Tucker type sufficient optimality condition for a feasible point to be conditionally properly efficient.

Theorem 2.2.1 (Kuhn-Tucker type Sufficient Conditions)

Consider the multiobjective problem (VP). Let there exist $\tau \in R^p$, $\lambda \in R^m$ such that (2.4)-(2.6) at a feasible point $x^* \in X_0$. If $\left(\tau_1 f_1, \ldots, \tau_p f_p\right)$ is $V-$ pseudo-invex and $\left(\lambda_1 g_1, \ldots, \lambda_m g_m\right)$ is $V-$ quasi-invex with respect to η. Then x^* is a conditionally properly efficient solution of (VP).

Proof: Let x be feasible for the problem (VP). Then, $g(x) \leq 0$. Since $\lambda_j g_j = 0$, $j = 1, \ldots, m$, then

$$\sum_{j=1}^{m} \lambda_j g_j(x) \leq \sum_{j=1}^{m} \lambda_j g_j\left(x^*\right).$$

Since $\beta_j\left(x, x^*\right) > 0$, $\forall\ j = 1, 2, \ldots, m$, we have

$$\sum_{j=1}^{m} \beta_j\left(x, x^*\right)\lambda_j g_j(x) \leq \sum_{j=1}^{m} \beta_j\left(x, x^*\right)\lambda_j g_j\left(x^*\right). \qquad (2.7)$$

Then by $V-$ quasi-invexity of $\left(\lambda_1 g_1, \ldots, \lambda_m g_m\right)$, we get

$$\sum_{j=1}^{m} \lambda_j g_j'\left(x^*\right)\eta\left(x, x^*\right) \leq 0.$$

Therefore, from (2.4) we have

$$\sum_{i=1}^{p} \tau_i f_i'\left(x^*\right)\eta\left(x, x^*\right) \geq 0.$$

Thus, from $V-$ pseudo-invexity of $\left(\tau_1 f_1, \ldots, \tau_p f_p\right)$, we have

$$\sum_{i=1}^{p} \alpha_i(x, x^*) \tau_i f_i'(x) \geq \sum_{i=1}^{p} \alpha_i(x, x^*) \tau_i f_i'(x^*).$$

Since $\alpha_i(x, x^*) > 0$, $\forall\, i = 1,..., p$ and $\tau > 0$, we have

$$f_i(x^*) \leq f_i(x), \quad \forall\, i = 1,..., p.$$

Thus, x^* is an efficient solution for (VP).

Now assume that x^* is not a conditionally properly efficient solution for (VP). Therefore, there exists a feasible for (VP) and an index such that for every positive function $M(x)$, we have $f_i(x^*) > M(x) f_j(x)$ for all j satisfying $f_j(x) > f_j(x^*)$ whenever $f_i(x) < f_i(x^*)$.

This means $f_i(x^*) - f_i(x)$ can be made arbitrarily large and hence for $\tau > 0$ and $\alpha_i(x, x^*) > 0$, $\forall\, i = 1,..., p$.

The inequality $\sum_{i=1}^{p} \alpha_i(x, x^*) \tau_i \left[f_i(x^*) - f_i(x) \right] > 0$ is obtained.

By V – pseudo-invexity of $\left(\tau_1 f_1, ..., \tau_p f_p \right)$, we get

$$\sum_{i=1}^{p} \tau_i f_i'(x^*) < 0. \tag{2.8}$$

Now from (2.8) and (2.4), we get

$$\sum_{j=1}^{m} \lambda_j g_j'(x^*) \eta(x, x^*) > 0. \tag{2.9}$$

By V – quasi-invexity of $\left(\lambda_1 g_1, ..., \lambda_m g_m \right)$ and (2.9), we get

$$\sum_{j=1}^{m} \beta_j(x, x^*) \lambda_j g_j(x) > \sum_{j=1}^{m} \beta_j(x, x^*) \lambda_j g_j(x^*),$$

which is a contradiction to (2.7).

Hence, x^* is a conditionally properly efficient solution of (VP).

2.3 Necessary and Sufficient Optimality Conditions for a Class of Nondifferentiable Multiobjective Programs

Mond (1974) considered a class of nondifferentiable mathematical programming problems of the form:

(NDP):   Minimize $f(x) + \left(x^T B x \right)^{\frac{1}{2}}$

$$\text{subject to } g(x) \geq 0, \tag{2.10}$$

where f and g are differentiable functions from R^n to R and R^m, respectively and B is an $n \times n$ positive semi-definite (symmetric) matrix.

Later Mond, Husain and Durga Prasad (1991) extended the work of Mond (1974) to multiobjective case:

$$\text{(NDVP): Minimize } \left(f_1(x) + \left(x^T B_1 x \right)^{\frac{1}{2}}, ..., f_p(x) + \left(x^T B_p x \right)^{\frac{1}{2}} \right)$$

$$\text{subject to } g(x) \geq 0. \tag{2.11}$$

In the subsequent analysis, we shall frequently use the following generalized Schwarz inequality [Riesz and Sz-Nagy (1955, pp. 262)]

$$x^T B z \leq \left(x^T B x \right)^{\frac{1}{2}} \left(z^T B z \right)^{\frac{1}{2}}, \quad \forall \ x, z \in R^n,$$

where B is an $n \times n$ positive semi-definite (symmetric) matrix.

We now state Kuhn-Tucker type necessary conditions.

Lemma 2.3.1. [Kuhn-Tucker type necessary condition]

Let x^* be an efficient solution of (NDVP). Then there exist $\tau \in R^p$, $\lambda \in R^m$ such that

$$\sum_{i=1}^{p} \tau_i \left[f_i'(x^*) + B_i z_i \right] + \sum_{j \in I(x^*)} \lambda_j g_j'(x^*) = 0 \tag{2.12}$$

$$\lambda_j g_j(x^*) = 0, \tag{2.13}$$

$$z^T B_i z_i \leq 1, \quad i = 1, ..., p \tag{2.14}$$

$$\left(x^{*T} B_i x^* \right)^{\frac{1}{2}} = x^{*T} B_i z_i, \tag{2.15}$$

$$\tau > 0, \ \lambda \geq 0, \ \sum_{i=1}^{p} \tau_i = 1, \tag{2.16}$$

where $I(x^*) = \{ j : g_j(x^*) = 0 \} \neq \phi$.

Theorem 2.3.2 [Sufficient Optimality Condition]

Let x^* be an efficient solution of (NDVP) and let there exist scalars $\tau \in R^p$ and λ such that

$$\sum_{i=1}^{p} \tau_i \left[\nabla_x f_i \left(x^* \right) + B_i z_i \right] + \sum_{j \in I(x^*)} \lambda_j \nabla_x g_j \left(x^* \right) = 0 \qquad (2.17)$$

$$\lambda_j g_j \left(x^* \right) = 0, \qquad (2.18)$$

$$z^T B_i z_i \leq 1, \qquad i = 1,...,p \qquad (2.19)$$

$$\left(x^{*T} B_i x^* \right)^{\frac{1}{2}} = x^{*T} B_i z_i, \qquad (2.20)$$

$$\tau > 0, \; \lambda \geq 0, \qquad (2.21)$$

where $I\left(x^* \right) = \left\{ j : g_j \left(x^* \right) = 0 \right\} \neq \phi$.

If $\left(\tau_1 \left(f_1 +{}^{\cdot T} B_1 z_1 \right),...,\tau_p \left(f_p +{}^T B_p z_p \right) \right)$ is $V-$ pseudo-invex and $\left(\lambda_1 g_1,...,\lambda_m g_m \right)$ is $V-$ quasi-invex with respect to the same η and for all piecewise smooth $z_i \in R^n$. Then x^* is conditionally properly efficient solution for (NDVP).

Proof: Let x be feasible for problem (NDVP). Then $x \in S$, $g(x) \leq 0$. Since $\lambda_j g_j \left(x^* \right) = 0$, $j = 1,...,m$, then

$$\sum_{j=1}^{m} \lambda_j g_j (x) \leq \sum_{j=1}^{m} \lambda_j g_j \left(x^* \right). \qquad (2.22)$$

Since $\beta_j \left(x, x^* \right) > 0$, $\forall j = 1,...,m$, we have

$$\sum_{j=1}^{m} \beta_j \left(x, x^* \right) \lambda_j g_j (x) \leq \sum_{j=1}^{m} \beta_j \left(x, x^* \right) \lambda_j g_j \left(x^* \right). \qquad (2.23)$$

Then by $V-$ pseudo-invexity of $\left(\lambda_1 g_1,...,\lambda_m g_m \right)$, we get

$$\sum_{j=1}^{m} \lambda_j \nabla_x g_j \left(x^* \right) \eta \left(x, x^* \right) \leq 0.$$

Therefore, from (2.12), we have

$$\sum_{i=1}^{p} \tau_i \left[\nabla_x f_i \left(x^* \right) + B_i z_i \right] \geq 0.$$

Thus, from $V-$ pseudo-invexity of $\left(\tau_1 \left(f_1 +{}^{\cdot T} B_1 z \right),...,\tau_p \left(f_p +{}^T B_p z \right) \right)$, we have

$$\sum_{i=1}^{p} \alpha_i\left(x,\, x^*\right)\tau_i\left[f_i(x) + x^T B_i z\right] \geq \sum_{i=1}^{p} \alpha_i\left(x,\, x^*\right)\tau_i\left[f_i\left(x^*\right) + x^{*^T} B_i z\right]. \quad (2.24)$$

That is

$$\alpha_i\left(x,\, x^*\right)\tau_i\left[f_i\left(x^*\right) + x^{*^T} B_i z\right] \leq \alpha_i\left(x,\, x^*\right)\tau_i\left[f_i(x) + x^T B_i z\right] \ \forall \ i,$$

and $\alpha_i(x, x^*)\tau_i\left[f_i(x^*) + x^{*^T} B_i z\right] < \alpha_j(x, x^*)\tau_j\left[f_j(x) + x^T B_j z\right]$ for at least one j.

Since, $\alpha_i\left(x,\, x^*\right) > 0 \ \forall \ i$ and $\tau > 0$, we get

$$f_i\left(x^*\right) + x^{*^T} B_i z \leq f_i(x) + x^T B_i z \ \forall \ i \ \text{ and}$$

$$f_j\left(x^*\right) + x^{*^T} B_j z < f_j(x) + x^T B_j z, \quad \text{for at least one } \ j.$$

Thus, x^* is an efficient solution of (NDVP).

Now assume that, x^* is not a conditionally properly efficient solution of (NDVP). Therefore, there exists $x \in K$ and an index i such that for every positive function $M(x)$, we have:

$$f_i\left(x^*\right) + \left(x^{*^T} B_i x^*\right)^{\frac{1}{2}} > M(x)\left(f_i\left(x^*\right) + \left(x^{*^T} B_i x^*\right)^{\frac{1}{2}} \right), \ \forall \ j \ \text{ satisfying}$$

$$f_j(x) + \left(x^T B_j x\right)^{\frac{1}{2}} > f_j\left(x^*\right) + \left(x^{*^T} B_j x^*\right)^{\frac{1}{2}},$$

whenever

$$f_i(x) + \left(x^T B_i x\right)^{\frac{1}{2}} < f_i\left(x^*\right) + \left(x^{*^T} B_i x^*\right)^{\frac{1}{2}}.$$

This means $f_i\left(x^*\right) + \left(x^{*^T} B_i x^*\right)^{\frac{1}{2}} - \left(f_i(x) + \left(x^T B_i x\right)^{\frac{1}{2}} \right)$ can be made

arbitrarily large and hence for $\tau > 0$ and $\alpha_i\left(x,\, x^*\right) > 0$, $\forall \ i$, the inequality

$$\sum_{i=1}^{p} \alpha_i\left(x,\, x^*\right)\tau_i\left(f_i\left(x^*\right) + \left(x^{*^T} B_i x^*\right)^{\frac{1}{2}} - f_i(x) - \left(x^T B_i x\right)^{\frac{1}{2}} \right) > 0$$

is obtained.

By V – pseudo-invexity of $\left(\tau_1\left(f_1 + \cdot^T B_1 \cdot\right), \ldots, \tau_p\left(f_p + \cdot^T B_p \cdot\right)\right)$, we get

$$\sum_{i=1}^{p} \tau_i\left(\nabla_x f_i\left(x^*\right) + B_i z\right)\eta\left(x,\, x^*\right) < 0. \quad\quad (2.25)$$

Now from (2.12) and (2.25), we get

$$\sum_{j \in I(x^*)} \lambda_j \nabla_x g_j(x^*) \eta(x, x^*) > 0.$$

By V – quasi-invexity of $(\lambda_1 g_1, ..., \lambda_m g_m)$, we have

$$\sum_{j \in I(x^*)} \beta_j(x, x^*) \lambda_j \nabla_x g_j(x) > \sum_{j \in I(x^*)} \beta_j(x, x^*) \lambda_j \nabla_x g_j(x^*), \qquad (2.26)$$

which is a contradiction to (2.23).

Hence, x^* is a conditionally properly efficient solution of (NDVP).

2.4 Duality

Several approaches to duality for the multiobjective optimization problem may be found in the literature. These include the use of vector valued Lagrangian, see for example Tanino and Sawaragi (1979), Weir (1987), White (1985) and Lagrangians incorporating matrix Lagrange multipliers, Bitran (1981), Corley (1981), Ivanov and Nehse (1985). Weir and Mond (1989) generalized the scalar duality results of Wolfe (1961), Mond and Weir (1981) and Bector and Bector (1987) to multiobjective optimization problem under the assumption of convexity. A vast number of works have appeared dealing with duality in multiobjective programs under different assumptions of convexity, for example, Preda (1992), Egudo (1989), Kaul, Suneja and Lalitha (1993), Rueda and Hanson (1988), Kaul, Suneja and Srivastava (1994) and Mond and Smart (1989) to mention a few.

Jeyakumar and Mond (1992) established the duality results for (VP) considered above in Section 2.1 under generalized V – invexity assumptions. The dual problem for (VP) is:

(VD): V – Maximize $(f_1(u), ..., f_p(u))$

subject to $\displaystyle\sum_{i=1}^{p} \tau_i f_i'(u) + \sum_{j=1}^{m} \lambda_j g_j'(u) = 0,$ (2.27)

$$\lambda_j g_j(u) \geq 0, \quad j = 1, ..., m, \qquad (2.28)$$

$$\tau \geq 0, \tau e = 1, \lambda \geq 0, \qquad (2.29)$$

where $e = (1, ..., 1) \in R^p$.

By considering the concept of weak minimum Jeyakumar and Mond (1992) demonstrated that V – pseudo-invexity of $(\tau_1 f_1, ..., \tau_p f_p)$ and

V – quasi-invexity of $(\lambda_1 g_1, ..., \lambda_m g_m)$ with respect to the same kernel function η was sufficient for weak duality to hold between the primal problem (VP) and its Mond-Weir type dual (VD) namely;

Theorem 2.4.1. (Weak Duality)
Consider the multiobjective problems (VP) and (VD). Let x be feasible for (VP) and let (u, τ, λ) be feasible for (VD). If $(\tau_1 f_1, ..., \tau_p f_p)$ is V – pseudo-invex and $(\lambda_1 g_1, ..., \lambda_m g_m)$ is V – quasi-invex with respect to the same η, then

$$\left(f_1(x), ..., f_p(x)\right)^T - \left(f_1(u), ..., f_p(u)\right)^T \notin -\text{int } R_+^p .$$

Mond and Weir (1981) proposed a number of different duals to the scalar valued minimization problem. Here we show that there are analogous results for the multiobjective optimization problem (VP) with generalized V – invexity assumptions.

Theorem 2.4.2 (Weak Duality)
If for all feasible (x, u, τ, λ)

(a). f_i, $i = 1, ..., p$ is V – pseudo-invex and $(\lambda_1 g_1, ..., \lambda_m g_m)$ is V – quasi-invex; or

(b). $(\tau_1 f_1, ..., \tau_p f_p)$ is V – pseudo-invex and $(\lambda_1 g_1, ..., \lambda_m g_m)$ is V – quasi-invex; or

(c). $(f_1, ..., f_p)$ is V – quasi-invex and $(\lambda_1 g_1, ..., \lambda_m g_m)$ is strictly V – pseudo-invex; or

(d). $(\tau_1 f_1, ..., \tau_p f_p)$ is V – quasi-invex and $(\lambda_1 g_1, ..., \lambda_m g_m)$ is strictly V – pseudo-invex, then $f(x) \not< f(u)$.

Proof:
(a). Assume contrary to the result, i. e., for x feasible for (VP) and (u, τ, λ) feasible for (VP), suppose $f_i(x) < f_i(u)$, for all $i = 1, ..., p$. Since $\alpha_i(x, u) > 0$, $\forall i = 1, ..., p$, we have

$$\alpha_i(x, u) f_i(x) < \alpha_i(x, u) f_i(u), \quad \forall \ i = 1, ..., p.$$

Therefore, $\sum_{i=1}^{p} \alpha_i(x, u) f_i(x) < \sum_{i=1}^{p} \alpha_i(x, u) f_i(u).$

By V – pseudo-invexity of f_i, $i = 1, ..., p$, we have

$$\sum_{i=1}^{p} f_i'(u)\eta(x, u) < 0.$$

Since $\tau \geq 0$,

$$\sum_{i=1}^{p} \tau_i f_i'(u)\eta(x, u) < 0. \tag{2.30}$$

Since $\lambda_j g_j(x) \leq \lambda_j g_j(u), \quad \forall \ j = 1,...,m.$

Again, since $\beta_j(x, u) > 0, \ \forall \ j = 1,...,m$, we have

$$\sum_{j=1}^{m} \beta_j(x, u)\lambda_j g_j(x) \leq \sum_{j=1}^{m} \beta_j(x, u)\lambda_j g_j(u).$$

Now, V − quasi-invexity implies that

$$\sum_{j=1}^{m} \lambda_j g_j'(u)\eta(x, u) \leq 0. \tag{2.31}$$

Combining (2.30) and (2.31), gives

$$\left(\sum_{i=1}^{p} \tau_i f_i'(u) + \sum_{j=1}^{m} \lambda_j g_j'(u)\right)\eta(x, u) < 0,$$

which contradicts the constraint (2.27) of (VD).

(b). Let x be feasible for (VP) and $f_i(x) < f_i(u), \ \forall \ i = 1,...,p$. Since $\tau \geq 0$ and $\alpha_i(x, u) > 0, \ \forall \ i = 1,...,p$, it follows that

$$\sum_{i=1}^{p} \alpha_i(x, u)\tau_i f_i(x) < \sum_{i=1}^{p} \alpha_i(x, u)\tau_i f_i(u),$$

and V − pseudo-invexity of $(\tau_1 f_1,...,\tau_p f_p)$ implies

$$\sum_{i=1}^{p} \tau_i f_i'(u)\eta(x, u) < 0.$$

Rest of the proof goes on the lines of the proof of part (a).

(c). Let x be feasible for (VP) and (u, τ, λ) feasible for (VD). Suppose $f_i(x) < f_i(u), i = 1,...,p$. Since $\alpha_i(x, u) > 0, \ \forall \ i = 1,...,p$, we have

$$\sum_{i=1}^{p} \alpha_i(x, u)f_i(x) < \sum_{i=1}^{p} \alpha_i(x, u)f_i(u).$$

The V − quasi-invexity of $(f_1,...,f_p)$ implies that

$$\sum_{i=1}^{p} f_i'(u)\eta(x, u) \leq 0.$$

Since $\tau \geq 0,$

$$\sum_{i=1}^{p} \tau_i f_i'(u)\eta(x,u) \leq 0.$$

By (2.27),

$$\sum_{j=1}^{m} \lambda_j g_j'(u)\eta(x,u) \geq 0.$$

Since $(\lambda_1 g_1,...,\lambda_m g_m)$ is strictly $V-$pseudo-invex, we have

$$\sum_{j=1}^{m} \beta_j(x,u)\lambda_j g_j(x) > \sum_{j=1}^{m} \beta_j(x,u)\lambda_j g_j(u),$$

which is a contradiction since $\lambda_j g_j(x) \leq 0$ and $\lambda_j g_j(u) \geq 0$ and $\beta_j(x,u) > 0, \ \forall \ j = 1,...,m.$

(d). Let x be feasible for (VP) and (u,τ,λ) feasible for (VD). Suppose $f_i(x) < f_i(u), i = 1,...,p.$ Since $\alpha_i(x,u) > 0, \ \forall \ i = 1,...,p,$ and $\tau \geq 0,$ we have

$$\sum_{i=1}^{p} \alpha_i(x,u)\tau_i f_i(x) < \sum_{i=1}^{p} \alpha_i(x,u)\tau_i f_i(u).$$

The $V-$quasi-invexity of $(\tau_1 f_1,...,\tau_p f_p)$ implies that

$$\sum_{i=1}^{p} \tau_i f_i'(u)\eta(x,u) \leq 0.$$

By (2.27)

$$\sum_{j=1}^{m} \lambda_j g_j'(u)\eta(x,u) \geq 0,$$

and since $(\lambda_1 g_1,...,\lambda_m g_m)$ is strictly $V-$pseudo-invex, we get

$$\sum_{j=1}^{m} \lambda_j g_j(x) > \sum_{j=1}^{m} \lambda_j g_j(u),$$

which is a contradiction since $\lambda_j g_j(x) \leq 0$ and $\lambda_j g_j(u) \geq 0$.

Theorem 2.4.3: If x^0 is feasible for (VP) and (u^0, τ^0, λ^0) is feasible for (VD) such that $f(x^0) = f(u^0)$ and for all feasible (u,τ, λ) of (VD), one of the conditions (a)-(d) hold. Then x^0 is conditionally properly efficient for (VP) and $\left(u^{0'}, \tau^0, \lambda^0\right)$ is conditionally properly efficient for (VD).

Proof: Suppose x^0 is not an efficient solution for (VP), then there exists x feasible for (VP) such that $f_i(x) \leq f_i(x^0) \ \forall \ i = 1,...,p$. Using the assumption $f_i(x^0) \leq f_i(u^0) \ \forall \ i = 1,...,p$, a contradiction to Theorem 2.4.2 is obtained. Hence x^0 is an efficient solution for (VP). Similarly it can be ensured that $\left(u^0, \tau^0, \lambda^0\right)$ is efficient solution for (VD).

Now suppose that x^0 is not conditionally properly efficient for (VP). Therefore, for every positive function $M(x) > 0$, there exists $\bar{x} \in X$ feasible for (VP) and an index i such that

$$f_i(x^0) - f_i(\bar{x}) > M(x)\left(f_j(\bar{x}) - f_j(x^0)\right) \quad \text{for all } j \text{ satisfying}$$

$$f_j(\bar{x}) > f_j(x^0), \text{ whenever } f_i(\bar{x}) < f_i(x^0).$$

This means $f_i(x^0) - f_i(\bar{x})$ can be made arbitrarily large and hence for $\tau^0 > 0$, the inequality

$$\sum_{i=1}^{p} \tau_i \left(f_i(x^0) - f_i(\bar{x})\right) > 0, \tag{2.32}$$

is obtained.

Now from feasibility conditions, we have

$$\lambda_j^0 g_j(\bar{x}) \leq \lambda_j^0 g_j(u^0), \ \forall \ j = 1,...,m.$$

Since $\beta_j(\bar{x}, u^0) > 0, \ \forall \ j = 1,...,m$

$$\sum_{j=1}^{m} \beta_j(\bar{x}, u^0) \lambda_j^0 g_j(\bar{x}) \leq \sum_{j=1}^{m} \beta_j(\bar{x}, u^0) \lambda_j^0 g_j(u^0).$$

By $V-$quasi-invexity of $\left(\lambda_1 g_1,...,\lambda_m g_m\right)$, we have

$$\sum_{j=1}^{m} \lambda_j^0 g_j(u^0) \eta(\bar{x}, u^0) \leq 0.$$

Therefore, from (2.27), we get

$$\sum_{i=1}^{p} \tau_i^0 f_i(u^0) \eta(\bar{x}, u^0) \geq 0.$$

Since $\tau \geq 0, \sum_{i=1}^{p} \tau_i = 1$, we have

$$\sum_{i=1}^{p} f_i(u^0) \eta(\overline{x}, u^0) \geq 0.$$

By V $-$ pseudo-invexity of f_i, $i = 1, ..., p$, we have

$$\sum_{i=1}^{p} \alpha_i(\overline{x}, u^0) f_i(\overline{x}) \geq \sum_{i=1}^{p} \alpha_i(\overline{x}, u^0) f_i(u^0). \tag{2.33}$$

On using the assumption $f(x^0) = f(u^0)$ in (2.33), we get

$$\sum_{i=1}^{p} \alpha_i(\overline{x}, u^0) f_i(\overline{x}) \geq \sum_{i=1}^{p} \alpha_i(\overline{x}, u^0) f_i(x^0).$$

Since $\alpha_i(\overline{x}, u^0) > 0$, $\forall i = 1, ..., p$ and $\tau_i^0 > 0$, $\forall i = 1, ..., p$, we get

$$\sum_{i=1}^{p} \tau_i^0 f_i(\overline{x}) \geq \sum_{i=1}^{p} \tau_i^0 f_i(x^0),$$

that is, $\sum_{i=1}^{p} \tau_i^0 \left[f_i(x^0) - f_i(\overline{x}) \right] \leq 0,$

which is a contradiction to (2.32).

Hence x^0 is a conditionally properly efficient solution for (VP).

We now suppose that (u^0, τ^0, λ^0) is not conditionally properly efficient solution for (VD). Therefore, for every positive function $M(x) > 0$, there exists a feasible $(\overline{u}, \overline{\tau}, \overline{\lambda})$ feasible for (VD) and an index i such that $f_i(\overline{u}) - f_i(u^0) > M(x)\left(f_i(u^0) - f_i(\overline{u}) \right)$ for all j satisfying $f_j(\overline{u}) > f_j(u^0)$, whenever $f_i(\overline{u}) < f_i(u^0)$.

This means $f_i(\overline{u}) - f_i(u^0)$ can be made arbitrarily large and hence for $\tau^0 > 0$, the inequality

$$\sum_{i=1}^{p} \tau_i^0 \left(f_i(\overline{u}) - f_i(u^0) \right) > 0, \tag{2.34}$$

is obtained.

Since x^0, (u^0, τ^0, λ^0) feasible for (VP) and (VD), respectively, it follows as in first part

$$\sum_{i=1}^{p} \tau_i^0 \left(f_i(\overline{u}) - f_i(u^0) \right) \leq 0,$$

which contradicts (2.34). Hence $\left(u^0, \tau^0, \lambda^0\right)$ is conditionally properly efficient solution for (VD).

Remark 2.4.2: In the proof of above Theorem we have only used generalized invexity conditions of part (a) of Theorem 2.4.2. Theorem 2.4.3 can be established for other V – invexity conditions mentioned in Theorem 2.4.2.

Theorem 2.4.4 (Strong Duality):

Let x^0 be efficient for (VP) and let one of (a)-(d) of Theorem 2.4.2 hold and the Kuhn-Tucker constraint qualification is satisfied. Then there exists (τ, λ) such that $\left(x^0, \tau, \lambda\right)$ is feasible for (VD) and the objective values of (VP) and (VD) are equal at x^0, and $\left(x^0, \tau, \lambda\right)$ is conditionally properly efficient for the problem (VD).

Proof: Since x^0 is an efficient solution for (VP) at which the Kuhn-Tucker type necessary conditions, there exists (τ, λ) such that $\left(x^0, \tau, \lambda\right)$ is feasible for (VD). Clearly the values of (VP) and (VD) are equal at x^0, since the objective functions for both problems are the same. The conditional proper efficiency of $\left(x^0, \tau, \lambda\right)$ for the problem (VD) follows from Theorem 2.4.3.

2.5 Duality for a Class of Nondifferentiable Multiobjective Programming

Mond (1974) considered a class of nondifferentiable mathematical programming problems of the form:

(P): $\qquad$ Minimize $f(x) + \left(x^T Bx\right)^{\frac{1}{2}}$

$\qquad\qquad$ subject to $g(x) \geq 0$,

where f and g are differentiable functions from R^n to R and R^m, respectively and B is an $n \times n$ positive semi-definite (symmetric) matrix. With the assumption that f is convex and g is concave, duality results were proved for a Wolfe type dual.

Mond and Smart (1989) weakened convexity requirements to invexity and its generalizations. Mond, Husain and Durga Prasad (1991) considered the following multiobjective nondifferentiable programming problem:

(NDVP):   Minimize $\left(f_1(x)+\left(x^T B_1 x\right)^{\frac{1}{2}} ,..., f_p(x)+\left(x^T B_p x\right)^{\frac{1}{2}} \right)$

subject to $g(x) \geq 0$,

and presented Mond-Weir type (1981) dual given below and established various duality results, viz., weak, strong and converse duality theorems under convex assumptions.

Lal et al. (1994) weakened convexity requirements to invexity and obtained weak duality theorem.

In relation to (NDVP) we associate the following dual nondifferentiable multiobjective maximization problem:

(NDVD): Maximize $\left(f_1(u)+\left(u^T B_1 z_1\right),..., f_p(u)+\left(u^T B_p z_p\right)\right)$

Subject to $\displaystyle\sum_{i=1}^{p} \tau_i \left[\nabla_x f_i(u)+ B_i z_i\right]+\sum_{j=1}^{m}\lambda_j \nabla_x g_j(u)=0,$ (2.35)

$z^T B_i z_i \leq 1,$ $i=1,...,p$ (2.36)

$\displaystyle\sum_{j=1}^{m}\lambda_j g_j(u) \geq 0,$ (2.37)

$\tau > 0,\ \lambda \geq 0,\ \displaystyle\sum_{i=1}^{p}\tau_i = 1.$ (2.38)

Let H denote the set of feasible solutions for (NDVD).

The following Theorem generalizes the weak duality theorem of Lal et al. (1994).

Theorem 2.5.1. (Weak Duality):
Let $x \in K$ and $\left(u,\tau,\lambda,z_1,...,z_p\right) \in H$ and

$$\left(\tau_1\left(f_1 +.^T B_1 z_1\right),...,\tau_p\left(f_p +.^T B_p z_p\right)\right)$$

is $V -$ pseudo-invex and $\left(\lambda_1 g_1,...,\lambda_m g_m\right)$ is $V -$ quasi-invex with respect to the same η and for all piecewise smooth $z_i \in R^n$. Then the following can not hold:

$$f_i(x)+\left(x^T B_i x\right)^{\frac{1}{2}} \leq f_i(u)+u^T B_i z_i,\quad \forall\ i=1,...,p$$

and $f_{i_0}(x)+\left(x^T B_{i_0} x\right)^{\frac{1}{2}} \leq f_{i_0}(u)+u^T B_{i_0} z_{i_0},$ for at least one i_0.

Proof: Let x be feasibility conditions

$$\lambda_j g_j(x) \le \lambda_j g_j(u), \quad j = 1, \ldots, m,$$

and $\lambda_{j_0} g_{j_0}(x) \le \lambda_{j_0} g_{j_0}(u)$, for atleast one $j \in \{1, \ldots, m\}$.

Since $\beta_j(x, u) > 0$, $\forall\, j = 1, \ldots, m$, we have

$$\sum_{j=1}^{m} \beta_j(x, u) \lambda_j g_j(x) \le \sum_{j=1}^{m} \beta_j(x, u) \lambda_j g_j(u).$$

Then by V – quasi-invexity of $(\lambda_1 g_1, \ldots, \lambda_m g_m)$, we get

$$\sum_{j=1}^{m} \lambda_j \nabla_x g_j(u) \eta(x, u) \le 0,$$

And so from (2.35), we have

$$\sum_{i=1}^{p} \tau_i [\nabla_x f_i(u) + B_i z_i] \eta(x, u) \ge 0.$$

Thus, from V – pseudo-invexity of
$$\left(\tau_1 \left(f_1 + {}^T B_1 z \right), \ldots, \tau_p \left(f_p + {}^T B_p z \right) \right),$$

we have

$$\sum_{i=1}^{p} \alpha_i(x, u) \tau_i [f_i(x) + x^T B_i z] \ge \sum_{i=1}^{p} \alpha_i(x, u) \tau_i [f_i(u) + u^T B_i z].$$

But $x^T B_i z_i \le \left(x^T B_i x \right)^{\frac{1}{2}} \left(z_i^T B_i z_i \right)^{\frac{1}{2}}$ (By Schwarz inequality) $\le \left(x^T B_i x \right)^{\frac{1}{2}}$ (by (2.14)).

Now, from (2.15) and (2.38), we have

$$\sum_{i=1}^{p} \alpha_i(x, u) \tau_i \left[f_i(x) + \left(x^T B_i x \right)^{\frac{1}{2}} \right] \ge \sum_{i=1}^{p} \alpha_i(x, u) \tau_i [f_i(u) + u^T B_i z].$$

That is,

$$\alpha_i(x, u) \tau_i \left[f_i(x) + \left(x^T B_i x \right)^{\frac{1}{2}} \right] \ge \alpha_i(x, u) \tau_i [f_i(u) + u^T B_i z_i] \quad \forall\, i, \text{ and}$$

$$\alpha_{i_0}(x, u) \tau_{i_0} \left[f_{i_0}(x) + (x^T B_{i_0} x)^{\frac{1}{2}} \right] \ge \alpha_{i_0}(x, u) \tau_{i_0} \left[f_{i_0}(u) + u^T B_{i_0} z_{i_0} \right], \text{.for at}$$

least one $i_0 \in \{1, \ldots, p\}$.

Since, $\alpha_i(x, u) > 0$ $\forall\, i$ and $\tau \ge 0$, we get

$$f_i(x) + \left(x^T B_i x \right)^{\frac{1}{2}} \ge f_i(u) + u^T B_i z \quad \forall\, i \quad \text{and}$$

$$f_{i_0}(x)+\left(x^T B_{i_0} x\right)^{\frac{1}{2}} > f_{i_0}(u)+u^T B_{j_0} z_{i_0}, \qquad \text{for at least one } i_0.$$

Thus, the following can not hold:

$$f_i(x)+\left(x^T B_i x\right)^{\frac{1}{2}} \le f_i(u)+u^T B_i z_i, \quad \forall \ i=1,...,p$$

and $f_{i_0}(x)+\left(x^T B_{i_0} x\right)^{\frac{1}{2}} \le f_{i_0}(u)+u^T B_{i_0} z_{i_0},$ for at least one i_0.

Theorem 2.5.2.

Let $x \in K$ and $\left(u, \tau, \lambda, z_1, ..., z_p\right) \in H$ and the $V-$pseudo-invexity and $V-$quasi-invexity conditions of Theorem 2.5.1 hold. If

$$u^T B_i u = u^T B_i z_i, \quad i=1,...,p, \tag{2.39}$$

and the objective values are equal, then x is conditionally properly efficient for (NDVP) and $\left(u, \tau, \lambda, z_1, ..., z_p\right)$ is conditionally properly efficient for (NDVD).

Proof: Suppose x is not an efficient solution for (NDVP), then there exists $x_0 \in K$ such that

$$f_i(x_0)+\left(x_0^T B_i x_0\right)^{\frac{1}{2}} \le f_i(u)+\left(x^T B_i x\right)^{\frac{1}{2}}, \quad \forall \ i=1,...,p$$

and $f_{i_0}(x_0)+\left(x_0^T B_{i_0} x_0\right)^{\frac{1}{2}} < f_{i_0}(x)+\left(x^T B_{i_0} x\right)^{\frac{1}{2}},$ for at least one i_0.

Using (2.15), we get

$$f_i(x_0)+\left(x_0^T B_i x_0\right)^{\frac{1}{2}} \le f_i(u)+u^T B_i z_i, \quad \forall \ i=1,...,p,$$

$$f_{i_0}(x_0)+\left(x_0^T B_{i_0} x_0\right)^{\frac{1}{2}} < f_{i_0}(x)+u^T B_{i_0} z_{i_0}, \quad \text{for at least one } i_0.$$

This is a contradiction of weak duality Theorem 2.5.1. Hence x is an efficient solution for (NDVP). Similarly it can be ensured that $\left(u, \tau, \lambda, z_1, ..., z_p\right)$ is an efficient solution of (NDVD).

Now suppose that x is not conditionally properly efficient of (NDVP). Therefore, for every positive function $M(x)>0$, there exists $x_0 \in X$ feasible for (NDVP) and an index i such that

$$f_i(x) + (x^T B_i x)^{\frac{1}{2}} - \left(f_i(x_0) + (x_0^T B_i x)^{\frac{1}{2}} \right)$$

$$> M(x) \left(f_i(x_0) + (x_0^T B_i x)^{\frac{1}{2}} - f_i(x) + (x^T B_i x)^{\frac{1}{2}} \right)$$

for all j satisfying

$$f_j(x_0) + \left(x_0^T B_j x_0 \right)^{\frac{1}{2}} > f_j(x) + \left(x^T B_j x \right)^{\frac{1}{2}},$$

whenever

$$f_i(x_0) + \left(x_0^T B_i x_0 \right)^{\frac{1}{2}} < f_i(x) + \left(x^T B_i x \right)^{\frac{1}{2}}.$$

This means $f_i(x) + \left(x^T B_i x \right)^{\frac{1}{2}} - \left(f_i(x_0) + \left(x_0^T B_i x_0 \right)^{\frac{1}{2}} \right)$ can be made ar-

bitrarily large and hence for $\tau > 0$, the inequality

$$\sum_{i=1}^{p} \tau_i \left(f_i(x) + \left(x^T B_i x \right)^{\frac{1}{2}} - \left(f_i(x_0) + \left(x_0^T B_i x_0 \right)^{\frac{1}{2}} \right) \right) > 0, \qquad (2.40)$$

is obtained.

Now from feasibility conditions, we have

$$\lambda_j g_j \left(x^0 \right) \le \lambda_j g_j(u), \ \forall \ j = 1, \dots, m.$$

Since $\beta_j \left(x, u^0 \right) > 0, \ \forall \ j = 1, \dots, m$

$$\sum_{j=1}^{m} \beta_j \left(x_0, u \right) \lambda_j g_j(x_0) \le \sum_{j=1}^{m} \beta_j \left(x_0, u \right) \lambda_j g_j(u).$$

By V – quasi-invexity of $\left(\lambda_1 g_1, \dots, \lambda_m g_m \right)$, we have

$$\sum_{j=1}^{m} \lambda_j \nabla_x g_j(u) \eta(x_0, u) \le 0.$$

Therefore, from (2.35), we get

$$\sum_{i=1}^{p} \tau_i^0 f_i \left(u^0 \right) \eta \left(\overline{x}, u^0 \right) \ge 0.$$

Since $\tau \ge 0$, $\sum_{i=1}^{p} \tau_i = 1$, we have

$$\sum_{i=1}^{p} \tau_i \left(\nabla_x f_i(u) + B_i z_i \right) \eta \left(\overline{x}, u^0 \right) \ge 0.$$

By using V – pseudo-invexity conditions, we have

$$\sum_{i=1}^{p} \alpha_i(x_0, u)\tau_i\left(f_i(x_0) + x_o^T B_i z_i\right) \geq \sum_{i=1}^{p} \alpha_i(x_0, u)\tau_i\left(f_i(u) + u^T B_i z_i\right).$$

Since $\alpha_i(x_0, u) > 0$, $\forall\, i = 1, ..., p$, we have

$$\sum_{i=1}^{p} \tau_i\left(f_i(x_0) + x_o^T B_i z_i\right) \geq \sum_{i=1}^{p} \tau_i\left(f_i(u) + u^T B_i z_i\right).$$

Since the objective values of (NDVP) and (NDVD) are equal, we have

$$\sum_{i=1}^{p} \tau_i\left(f_i(x_0) + \left(x_o^T B_i x_0\right)^{\frac{1}{2}}\right) \geq \sum_{i=1}^{p} \tau_i\left(f_i(x) + \left(x^T B_i x\right)^{\frac{1}{2}}\right).$$

This yields

$$\sum_{i=1}^{p} \tau_i\left\{\left(f_i(x) + \left(x^T B_i x\right)^{\frac{1}{2}}\right) - \left(f_i(x_0) + \left(x_o^T B_i x_0\right)^{\frac{1}{2}}\right)\right\} \leq 0,$$

which is a contradiction to (2.40).

Hence x is a conditionally properly efficient solution for (NDVP).

We now suppose that $(u, \tau, \lambda, z_1, ..., z_p)$ is not conditionally properly efficient solution for (NDVD). Therefore, for every positive function $M(x) > 0$, there exists a feasible $(u_0, \tau_0, \lambda_0, z_1^0, ..., z_p^0)$ feasible for (NDVD) and an index i such that

$$f_i(u_0) + u_0^T B_i z_i^0 - \left(f_i(u) + u^T B_i z_i\right)$$
$$> M(x)\left(f_i(u) + u^T B_i z_i - f_i(u_0) - u_0^T B_i z_i^0\right)$$

for all j satisfying

$$f_j(u_0) + u_0^T B_j z_j^0 < f_j(u) + u^T B_j z_j,$$

whenever

$$f_i(u_0) + u_0^T B_i z_i^0 > f_i(u) + u^T B_i z_i.$$

This means $f_i(u_0) + u_0^T B_i z_i^0 - f_i(u) - u^T B_i z_i$ can be made arbitrarily large and hence for $\tau > 0$, the inequality

$$\sum_{i=1}^{p} \tau_i\left(f_i(u_0) + u_0^T B_i z_i^0 - f_i(u) - u^T B_i z_i\right) > 0, \qquad (2.41)$$

is obtained.

Since x, $(u, \tau, \lambda, z_1, ..., z_p)$ feasible for (NDVP) and (NDVD), respectively, it follows as in first part

$$\sum_{i=1}^{p} \tau_i \left(f_i(u_0) + u_0^T B_i z_i^0 - f_i(u) - u^T B_i z_i \right) \le 0,$$

which contradicts (2.41). Hence $\left(u, \tau, \lambda, z_1, ..., z_p \right)$ is conditionally properly efficient solution for (NDVD).

Theorem 2.5.3 (Strong Duality):
Let x be a conditionally properly efficient solution for (NDVP) at which a suitable constraint qualification is satisfied. Let the V − pseudo-invexity and V − quasi-invexity conditions of Theorem 2.5.1 be satisfied. Then there exists $\left(\tau, \lambda, z_1, ..., z_p \right)$ such that $\left(x = u, \tau, \lambda, z_1, ..., z_p \right)$ is a conditionally properly efficient solution for (NDVD) and

$$f_i(x) + \left(x^T B_i x \right)^{\frac{1}{2}} = f_i(u) + u^T B_i z_i, \quad i = 1, ..., p.$$

Proof: Since x is conditionally properly efficient solution for (NDVP) and a constraint qualification is satisfied at x, from the Kuhn-Tucker necessary condition Lemma 2.3.1, there exists $\left(\tau, \lambda, z_1, ..., z_p \right)$ such that $\left(x, \tau, \lambda, z_1, ..., z_p \right)$ is feasible for (NDVD). Since

$$\left(x^T B_i x \right)^{\frac{1}{2}} = x^T B_i z_i, \quad i = 1, ..., p,$$

the values of (NDVP) and (NDVD) are equal at x. By Theorem 2.5.2, $\left(x = u, \tau, \lambda, z_1, ..., z_p \right)$ is conditionally properly efficient solution of (NDVD).

2.6 Vector Valued Infinite Game and Multiobjective Programming

Karlin (1959) observed that matrix games were equivalent to a dual pair of linear programs, see also Charnes (1953) and Cottle (1963). More recently, Kawaguchi and Maruyama (1976) formulated dual linear programs corresponding to the linearly constrained matrix game. Kawaguchi and Maruyama (1976) considered a linearly constrained matrix game and using saddle point theory established an equivalence between this game and a pair of mutually dual linear programming problems.

Corley (1985) considered a two-person bi-matrix vector valued game in which strategy spaces are mixed and introduced the concept of solution of this game. He also established the necessary and sufficient conditions for the solution of such a game.

Chandra and Durga Prasad (1992) considered a constrained two-person zero-sum game with vector pay-off and discussed its relation with a pair of multiobjective programming problems. Consider the following two multiobjective programming problems (P) and (D):

$$
\text{(P):} \quad V - Min \begin{pmatrix} K_1(x, y) - x^T \left(\displaystyle\sum_{i=1}^{p} \tau_i \nabla_1 K_i(x, y) \right), \dots, \\[3mm] K_p(x, y) - x^T \left(\displaystyle\sum_{i=1}^{p} \tau_i \nabla_1 K_i(x, y) \right) \end{pmatrix}
$$

subject to $\displaystyle\sum_{i=1}^{p} \tau_i \nabla_1 K_i(x,\, y) \le 0,$ \hfill (2.42)

$$x \ge 0, \quad y \ge 0, \quad \tau \in \Lambda. \hfill (2.43)$$

$$
\text{(D):} \quad V - Max \begin{pmatrix} K_1(u, v) - x^T \left(\displaystyle\sum_{i=1}^{p} \mu_i \nabla_2 K_i(u, v) \right), \dots, \\[3mm] K_p(u, v) - x^T \left(\displaystyle\sum_{i=1}^{p} \mu_i \nabla_2 K_i(u, v) \right) \end{pmatrix}
$$

subject to $\displaystyle\sum_{i=1}^{p} \mu_i \nabla_2 K_i(u,\, v) \ge 0,$ \hfill (2.44)

$$u \ge 0, \quad v \ge 0, \quad \mu \in \Lambda, \hfill (2.45)$$

where $x, u \in R^m$; $y, v \in R^n$; $\tau, \mu \in R^p$; and $K : R^m \times R^n \to R^p$.

Corresponding to the multiobjective programming problems (P) and (D) as defined above, consider the following vector-valued infinite game $VG: \{S, T, K\}$, where,

(i) $S = \{x \in R^m : x \ge 0\}$ is the strategy space for player I,

(ii) $T = \{y \in R^n : y \ge 0\}$ is the strategy space for player II, and

(iii) $K : S \times T \to R^p$ defined by $K(x,\, y)$ is the pay-off to player I. The pay-off to player II will be taken as $K(y,\, x)$

In order to establish necessary and sufficient conditions we need the following definitions:

Definition 2.6.1 (Corley (1985)): A point $(\bar{x}, \bar{y}) \in S \times T$ is said to be an equilibrium point of the game G if

$K(x, \bar{y}) \not\geq K(\bar{x}, \bar{y}), \quad \forall \ x \in S$, and

$K(\bar{x}, y) \not\leq K(\bar{x}, \bar{y}), \quad \forall \ y \in T$.

Definition 2.6.2 (Tanino, Nakayama and Sawaragi (1985)):

Let $f : R^n \rightarrow R^p$. A point $\bar{x} \in S$, is said to be an efficient solution of the vector maximization problem. $V - \max \ f(x)$ over $x \in S$, if there does not exist any $x \in X$ such that $f(x) \geq f(\bar{x})$.

Definition 2.6.3 (Rodder (1977)): A point $(x^0, y^0) \in S \times T$ is called a solution of the max-min problem if

(i) y^0 is an efficient solution of $V - \min \ K(x^0, y), \ y \in T$.

(ii) $K(x^0, y^0) \not\leq K(x, y), \quad \forall \ x \in S$ and $y \in T$.

Definition 2.6.4 (Rodder (1977)): A point $(x^0, y^0) \in S \times T$ is called a solution of the min-max problem if

(i) x^0 is an efficient solution of $V - \max \ K(x, y^0), \ x \in S$.

(ii) $K(x^0, y^0) \not\geq K(x, y), \quad \forall \ x \in S$ and $y \in T$.

Definition 2.6.5 (Rodder (1977)): A point $(x^0, y^0) \in S \times T$ is called a generalized saddle point (x^0, y^0) solves both max-min and min-max problems.

Definition 2.6.6 (Rodder (1977)): The following statements are equivalent:

(i) (x^0, y^0) is a generalized saddle point of $K(x, y)$ in $S \times T$,

(ii) y^0 solves $V - \min \ K(x^0, y)$ and x^0 solves $V - \max \ K(x, y^0), \ y \in T$,

(iii) $K(x, y^0) \not\geq K(x^0, y^0), \forall \ x \in S$ and $K(x^0, y) \not\leq K(x^0, y^0), \forall \ y \in T$.

Chandra and Durga Prasad (1993) established the following necessary conditions:

If (x, y) is an equilibrium point of the game VG. Then there exists $\bar{\tau} \in R_+^p, \bar{\tau} \neq 0$ and $\bar{\mu} \in R_+^p, \bar{\mu} \neq 0$ such that $(\bar{x}, \bar{y}, \bar{\tau})$ and $(\bar{x}, \bar{y}, \bar{\mu})$ are efficient to multiobjective programming problems (P) and (D) respectively.

Sufficient conditions are also established under concave-convex assumption on K_i in Chandra and Durga Prasad (1993). The following Theorem is obtained under weaker convexity assumptions on K_i.

Theorem 2.6.1 (Sufficient Conditions): Let $(\bar{x}, \bar{y}, \bar{\tau})$ and $(\bar{x}, \bar{y}, \bar{\mu})$ be feasible for (P) and (D) respectively with

$$\bar{x}^T \sum_{i=1}^{p} \bar{\tau}_i \nabla_1 K_i(\bar{x}, \bar{y}) = 0 = \bar{y}^T \sum_{i=1}^{p} \bar{\mu}_i \nabla_2 K_i(\bar{x}, \bar{y})$$

and is an equilibrium point of the game $\bar{\tau} > 0$, $\bar{\mu} > 0$. Also let, for each $i = 1,...,p$, K_i be $V-$ incave-invex. Then $(\bar{x}, \bar{y})$ is an equilibrium point of the game VG.

Proof: We have to prove that

$K(\bar{x}, \bar{y}) \nleq K(x, \bar{y})$, $\forall x \in S$, and $K(\bar{x}, \bar{y}) \ngeq K(x, \bar{y})$, $\forall y \in T$.

If possible, let $K(\bar{x}, \bar{y}) \leq K(\hat{x}, \bar{y})$, for some $\hat{x} \in S$. Therefore,

$$\sum_{i=1}^{p} \tau_i K_i(\bar{x}, y) < \sum_{i=1}^{p} \tau_i K_i(\bar{x}, y).$$

Now by $V-$ incavity of $\tau_i K_i$, $i = 1,...,p$, at x, we have

$$\sum_{i=1}^{p} \alpha_i(\hat{x}, x)\tau_i \nabla_1 K_i(\bar{x}, \bar{y})\eta(\hat{x}, \bar{y}) > 0.$$

Since $\alpha_i(\hat{x}, x) > 0$, $\forall i = 1,...,p$, we have

$$\sum_{i=1}^{p} \tau_i \nabla_1 K_i(\bar{x}, \bar{y})\eta(\hat{x}, \bar{y}) > 0.$$

Since x is feasible, $\eta(\hat{x}, \bar{x}) + \bar{x} \geq 0 \Rightarrow \eta(\hat{x}, \bar{x}) = x - \bar{x}$ for some $x \geq 0$, we get

$$(x - \bar{x})^T \sum_{i=1}^{p} \tau_i \nabla_1 K_i(\bar{x}, \bar{y}) > 0,$$

that is,

$$x^T \sum_{i=1}^{p} \tau_i \nabla_1 K_i(\bar{x}, \bar{y}) > \bar{x}^T \sum_{i=1}^{p} \tau_i \nabla_1 K_i(\bar{x}, \bar{y}). \tag{2.46}$$

But (2.44) together with the hypothesis of the theorem yields

$$x^T \sum_{i=1}^{p} \tau_i \nabla_1 K_i(\bar{x}, \bar{y}) > 0,$$

which contradicts (2.42). Hence $K(\bar{x}, \bar{y}) \nleq K(x, \bar{y})$, $\forall\, x \in S$. Similarly we can show that $K(\bar{x}, \bar{y}) \ngeq K(\bar{x}, y)$, $\forall\, y \in T$.

Chapter 3: Multiobjective Fractional Programming

3.1 Introduction

Numerous decision problems in management science and problems in economic theory give rise to constrained optimization of linear or nonlinear functions. If in the nonlinear case the objective function is a ratio of two functions or involves several such ratios, then the optimization problem is called a fractional program.

Apart from isolated earlier results, most of the work in fractional programming were done since about 1960. The analysis of fractional programs with only one ratio has largely dominated the literature until about 1980. Since the first international conference with an emphasis on fractional programming the NATO advanced Study Institute on "Generalized Concavity in Optimization and Economics" (Schaible and Ziemba (1981)), that indicates a shift of interest from the single to the multiobjective case, see Singh and Dass (1989), Cambini, Castagnoli, Martein, Mazzoleni and Schaible (1990), Komlosi, Rapcsak and Schaible (1994), Mazzoleni (1992). It is interesting to note that some of the earliest publications in fractional programming, though not under this name, Von Neuman's classical paper on a model fo a general economic equilibrium [Von Neumann (1937)] analysis a multiobjective fractional program. Even a duality theory was proposed for this nonconcave program, and this at a time when linear programming hardly existed. However, this early paper was followed almost exclusively by articles in single objective fractional programming until the early 1980s.

Weir (1982) considered a multiobjective fractional programming problem with same denominators. Since then a great deal of work has been done with convexity and generalized convexity assumptions on the functions. Some of the contributions are by Singh (1986), Egudo (1988), Weir (1986, 1989), Kaul and Lyall (1989), Suneja and Gupta (1990), Mukherjee (1991), Singh and Hanson (1991), Preda (1992), Suneja and Lalitha (1993), Kaul, Suneja and Lalitha (1993), Suneja and Srivastava (1994) and Mishra and Mukherjee (1996a).

Throughout this chapter (except sections 3.4 and 3.5) we consider the following multiobjective fractional programming problem:

(MFP)   Minimize $\left(\dfrac{f_1(x)}{g_1(x)},...,\dfrac{f_p(x)}{g_p(x)} \right)$

subject to $h_j(x) \le 0, \quad j = 1,...,m, \ x \in X,$

where $f_i, g_i : X \to R, \ i = 1,...,p,$ and $h_j : X \to R, \ j = 1,...,m$ and differentiable functions, $f_i(x) \ge 0, \ g_i(x) > 0, \ i = 1,...,p, \ \forall \ x \in X$, and minimization entails obtaining efficient solutions properly efficient solutions/conditionally properly efficient solutions.

We consider the following parametric multiobjective problem $(FP)_{V'}$ for each $v \in R_+^p$, where R_+^p denotes the positive orthant of R^p.

$(FP)_{V'}$ Minimize $\left(f_1(x) - v_1 g_1(x),..., f_p(x) - v_p g_p(x) \right)$

subject to $h_j(x) \le 0, \quad j = 1,...,m, \ x \in X.$

The following lemma from Singh and Hanson (1991) connects the conditionally properly efficient solutions of (FP) and $(FP)_{V'}$.

Lemma 3.1.1 [Singh and Hanson (1991)]: Let x^* be conditionally properly efficient solution of (FP). Then there exists $v^* \in R_+^p$ such that x^* is conditionally properly efficient solution of $(FP)_{V^*}$. Conversely, if x^* is conditionally properly efficient solution of $(FP)_{V^*}$ where

$$v_i^* = \frac{f_i\left(x^* \right)}{g_i\left(x^* \right)}, \quad i = 1, 2,..., p,$$

then x^* is conditionally properly efficient solution for (FP).

We consider on the lines of Geoffrion (1968), the following scalar programming problem corresponding to $(FP)_{V^*}$:

$(MFP)_{v^*}^\tau$ Minimize $\displaystyle\sum_{i=1}^{p} \tau_i \left(f_i(x) - v_i^* g_i(x) \right)$

subject to $h_j(x) \le 0, \quad j = 1,...,m, \ x \in X.$

Then we have the following result from Singh and Hanson (1991):

Lemma 3.1.2 [Singh and Hanson (1991)]: If x^* is an optimal solution of $(MFP)_{v^*}^\tau$ for some $\tau \in R^p$ with strictly positive components where

$$v_i^* = \frac{f_i(x^*)}{g_i(x^*)}, \quad i = 1, 2, ..., p,$$

Then x^* is conditionally properly efficient solution of (MFP).

3.2 Necessary and Sufficient Conditions for Optimality

Let $x^* \in X$ be an efficient solution for (MFP). Then there exist $\tau^*, v^* \in R^p$ and $\lambda^* \in R^m$ such that

$$\sum_{i=1}^{p} \tau^* \left(\nabla f_i(x^*) - v_i^* \nabla g_i(x^*) \right) + \sum_{j=1}^{m} \lambda_j^* \nabla h_j(x^*) = 0 , \tag{3.1}$$

$$\lambda^{*T} h(x^*) = 0, \tag{3.2}$$

$$f_i(x^*) - v_i^* g_i(x^*) = 0, \quad i = 1, ..., p \tag{3.3}$$

$$v_p^* \geq 0, \quad (\tau^*, \lambda^*) \geq 0 . \tag{3.4}$$

Whenever we assume a constraint qualification for (MFP), we mean that (MFP) satisfies the Kuhn-Tucker constraint qualification or the weak Arrow-Hurwicz-Uzawa constraint qualification (Mangasarian (1969), p. 102). Kuhn-Tucker type necessary conditions are as follows:

For $x^* \in X$ an efficient solution for (MFP) and (MFP) satisfies a constraint qualification at x^*. Then there exist $\tau^*, v^* \in R^p$ and $\lambda^* \in R^m$ such that

$$\sum_{i=1}^{p} \tau^* \left(\nabla f_i(x^*) - v_i^* \nabla g_i(x^*) \right) + \sum_{j=1}^{m} \lambda_j^* \nabla h_j(x^*) = 0, \tag{3.5}$$

$$\lambda^{*T} h(x^*) = 0 \tag{3.6}$$

$$f_i(x^*) - v_i^* g_i(x^*) = 0, \quad \forall \ i = 1, ..., p, \tag{3.7}$$

$$\tau^*, v^*, w^* \geq 0, \quad \sum_{i=1}^{p} \tau_i^* = 1. \tag{3.8}$$

The following necessary optimality criteria for a feasible point x^* of (MFP) to be conditionally properly efficient can be proved on similar lines as that of Theorem 2 of Weir (1988).

Theorem 3.2.1: Let x^* be a conditionally properly efficient solution for (MFP). Assume that there exists $\bar{x} \in X$ such that $h_j(\bar{x}) < 0$, for $j = 1, \ldots, m$ and for $j \in I(x^*) = \{j : h_j(\bar{x}) = 0\}$ any one of the following conditions holds

(i) h_j is $V - \text{invex}$

(ii) h_j is $V - \text{pseudo-invex}$

on X with respect to η and $\alpha_i > 0$, $i = 1, \ldots, p$. Then there exist scalars $\tau_i^* > 0$, $i = 1, \ldots, p$, $\lambda_i^* \geq 0$, $i \in I(x^*)$ such that

$$\sum_{i=1}^{p} \tau_i^* \nabla\left(\frac{f_i(x^*)}{g_i(x^*)}\right) + \sum_{i \in I(x^*)} \lambda_i^* \nabla h_i(x^*) = 0 . \tag{3.9}$$

Proof: Since x^* is conditionally properly efficient for (MFP) therefore by Lemma 3.1.1 there exists $v^* \in R_+^p$ such that x^* is conditionally properly efficient for $(\text{MFP})_{v^*}$ where $v_i^* = \dfrac{f_i(x^*)}{g_i(x^*)}$, $i = 1, \ldots, p$. Since each h_j, $j = 1, \ldots, m$ satisfies (i) or (ii), there by proceeding on the same lines as in Theorem 2 of Weir (1988) we shall get the required result.

The following example verifies the above theorem for a multiobjective fractional programming problem with $p = m = 2$.

Example 3.2.1: Consider the following multiobjective fractional programming problem:

(FP1)   Minimize $\left(\dfrac{f_1(x)}{g_1(x)}, \dfrac{f_2(x)}{g_2(x)}\right)$

subject to $h_j(x) \leq 0$, $j = 1, 2$

where functions f_1, f_2, g_1, g_2, h_1 and h_2 are defined on $X = (-2, 2)$ as follows:

The feasible region is the closed interval $[0, 1]$. We observed that $x^* = 1$ is an efficient solution of (FP1) because for any feasible solution x of (FP1)

$$\frac{f_1(x)}{g_1(x)} - \frac{f_1(x^*)}{g_1(x^*)} = \frac{x^2 - 1}{3(x^2 + 2)} \leq 0$$

$$\frac{f_2(x)}{g_2(x)} - \frac{f_2\left(x^*\right)}{g_2\left(x^*\right)} = \frac{5(1-x)}{3(x+2)} \geq 0,$$

and if $\dfrac{f_1(x)}{g_1(x)} < \dfrac{f_1\left(x^*\right)}{g_1\left(x^*\right)}$,

Then $x < 1$ for which $\dfrac{f_2(x)}{g_2(x)} > \dfrac{f_2\left(x^*\right)}{g_2\left(x^*\right)}$. Now we will prove that

$x^* = 1$ is a conditionally properly efficient solution of (FP1).

For $x < 1$

$$\left(\frac{f_1\left(x^*\right)}{g_1\left(x^*\right)} - \frac{f_1(x)}{g_1(x)}\right) \bigg/ \left(\frac{f_2(x)}{g_2(x)} - \frac{f_2\left(x^*\right)}{g_2\left(x^*\right)}\right) = \frac{x^2 + 3x + 2}{5\left(x^2 + 2\right)}$$

is a function which attains a maximum value at $n = \sqrt{2}$ with value being $\left(4 + 3\sqrt{2}\right)$. Thus choosing $M\left(x^*\right) = x^2 + \dfrac{1}{2}$, it follows that $x^* = 1$ is a conditionally properly efficient solution of (FP1).

Now h_1 is the only constraint for which $h_1\left(x^*\right) = 0$. Define η, α_i, $i = 1,2$ and β_j, $j = 1,2$ by $\eta(x, u) = \dfrac{x - 2u}{2}$, $\alpha_i(x,u) = \dfrac{u}{2}$ $i = 1,2$ and $\beta_j(x,u) = 1, j = 1,2$. h_j is V − pseudo-invex with respect to η and β_j. Moreover, $\bar{x} = \dfrac{1}{2}$ is such that $h_1(\bar{x}) < 0$, $h_2(\bar{x}) < 0$. Thus, by Theorem 3.2.1, there exist $\tau_i^* > 0$, $\lambda_j^* \geq 0$ such that

$$\sum_{i=1}^{2} \tau_i^* \nabla\left(\frac{f_i\left(x^*\right)}{g_i\left(x^*\right)}\right) + \lambda_1^* \nabla h_1\left(x^*\right) = 0.$$

Clearly, $\tau_1^* = 1$, $\tau_2^* = 4$, $\lambda_1^* = 1$ satisfies the above equation.

We now give a number of sufficient optimality criteria for a feasible point x^* of (MFP) to be conditionally properly efficient for (MFP) under the assumptions of V − invexity and its generalizations:

Theorem 3.2.2: Suppose that there exists a feasible x^* for (MFP) and scalar $\tau_i^* > 0$, $i = 1,...,p$, $\lambda_i^* = 0$, $i \in I\left(x^*\right)$ such that

$$\sum_{i=1}^{p} \tau_i^* \left(\nabla f_i \left(x^* \right) - v_i^* \nabla g_i \left(x^* \right) \right) + \sum_{i \in I\left(x^* \right)} \lambda_i^* \nabla h_i \left(x^* \right) = 0$$

Where
$$v_i^* = \frac{f_i\left(x^* \right)}{g_i\left(x^* \right)}, \quad i = 1, \ldots, p$$

and $I\left(x^* \right) = \left\{ j : h_i \left(x^* \right) = 0 \right\} \neq \phi$. Then, if each $\left(f_i - v_i g_i \right)$, $i = 1, \ldots, p$ is $V - $ invex and h_j, $j \in I\left(x^* \right)$ is $V - $ invex with respect to the same η and α_i, $i = 1, \ldots, p$, β_i, $i \in I\left(x^* \right)$, then x^* is conditionally properly efficient solution for (MFP).

Proof: Since each $\left(f_i - v_i g_i \right)$, $i = 1, \ldots, p$ and h_j, $j \in I\left(x^* \right)$ are $V - $ invex with respect to the same η and

$$\alpha_i, \ i = 1, \ldots, p, \ \beta_i, \ i \in I\left(x^* \right),$$

and

$$\tau_i^* > 0, i = 1, \ldots, p, \ \lambda_i^* = 0, i \in I\left(x^* \right), \ v_i^* = \frac{f_i\left(x^* \right)}{g_i\left(x^* \right)}, \quad i = 1, \ldots, p,$$

we have

$$\sum_{i=1}^{m} \tau_i^* \left(f_i(x) - v_i^* g_i(x) \right) - \sum_{i=1}^{p} \tau_i^* \left(f_i\left(x^* \right) - v_i^* g_i\left(x^* \right) \right)$$

$$\geq \sum_{i=1}^{m} \tau_i^* \alpha_i\left(x, x^* \right) \left(\nabla f_i\left(x^* \right) - v_i^* \nabla g_i\left(x^* \right) \right) \eta\left(x, x^* \right)$$

$$= - \sum_{i \in I\left(x^* \right)} \lambda_i^* \beta_i\left(x, x^* \right) \nabla h_i\left(x^* \right) \eta\left(x, x^* \right)$$

$$\geq \sum_{i \in I\left(x^* \right)} \lambda_i^* \left(h_i\left(x^* \right) - h_i(x) \right)$$

$$\geq \sum_{i \in I\left(x^* \right)} - \lambda_i^* h_i(x) \quad \text{(Since } h_i\left(x^* \right) = 0, \ i \in I\left(x^* \right) \text{)}$$

$$\geq 0.$$

Therefore,

$$\sum_{i=1}^{m} \tau_i^* \left(f_i(x) - v_i^* g_i(x) \right) - \sum_{i=1}^{p} \tau_i^* \left(f_i\left(x^* \right) - v_i^* g_i\left(x^* \right) \right) \geq 0, \ \forall \ x \in X.$$

This implies that x^* minimizes $\sum_{i=1}^{m} \tau_i^* \left(f_i(x) - v_i^* g_i(x) \right)$ subject to $h_j(x) \leq 0$, $j = 1, \ldots, m$. Hence x^* is an optimal solution for $(\mathrm{FP})_{v^*}^{\tau}$.

Therefore, x^* is conditionally properly efficient solution for (MFP) due to Lemma 3.1.2.

Theorem 3.2.3 Suppose there exists a feasible x^* for (MFP) and scalar $\tau_i^* > 0$, $i = 1, \ldots, p$, $\lambda_i^* = 0$, $i \in I(x^*)$ such that (3.1)-(3.4) is satisfied. Then, if $I(x^*) = \phi$, $(f_i - v_i g_i)$, $i = 1, \ldots, p$ and h_j, $j \in I(x^*)$ are V − quasi-invex with respect to the same η, then x^* is conditionally properly efficient solution for (MFP).

Proof: Since for $h_i(x^*) = 0$, for $i \in I(x^*)$ and $h_i(x) \leq 0$, $i = 1, \ldots, m$, we have

$$h_i(x) - h_i(x^*) \leq 0, \quad i = 1, \ldots, m.$$

Since $\lambda_i \geq 0$, $i \in I(x^*)$, we have

$$\sum_{i \in I(x^*)} \lambda_i^* \left(h_i(x) - h_i(x^*) \right) \leq 0.$$

Now, by V − quasi-invexity of h_j, $j \in I(x^*)$, we have

$$\sum_{i \in I(x^*)} \lambda_i^* \beta_i(x, x^*) \nabla h_i(x^*) \eta(x, x^*) \leq 0.$$

On using the above inequality in (3.9), we obtain

$$\sum_{i=1}^{m} \tau_i^* \left(\nabla f_i(x^*) - v_i^* \nabla g_i(x^*) \right) \eta(x, x^*) \geq 0, \quad \forall\, x \in X.$$

Since $\alpha_i(x, x^*) > 0$, $i = 1, \ldots, p$, we have

$$\sum_{i=1}^{m} \tau_i^* \alpha_i(x, x^*) \left(\nabla f_i(x^*) - v_i^* \nabla g_i(x^*) \right) \eta(x, x^*) \geq 0.$$

Now, by V − pseudo-invexity of $(f_i - v_i g_i)$, $i = 1, \ldots, p$, we have

$$\sum_{i=1}^{m} \tau_i^* \left(f_i(x) - v_i^* g_i(x) \right) - \sum_{i=1}^{p} \tau_i^* \left(f_i(x^*) - v_i^* g_i(x^*) \right) \geq 0, \quad \forall\, x \in X.$$

Thus x^* is an optimal solution of $(\mathrm{FP})_{v^*}^{\tau}$ for τ^* with strictly positive components. Hence, by Lemma 3.1.2, x^* is conditionally properly efficient for (MFP).

Theorem 3.2.4: Suppose that there exists a feasible x^* for (MFP) and scalar $\tau_i^* > 0$, $i = 1,..., p$, $\lambda_i^* = 0$, $i \in I(x^*)$ such that (3.1)-(3.4) is satisfied. Then, if $I(x^*) = \phi$, $\sum_{i=1}^{p} \tau_i^*(f_i - v_i g_i)$ is $V -$ quasi-invex and $\lambda_i^* h_i$, $i \in I(x^*)$ are $V -$ strictly pseudo-invex with respect to the same η, then x^* is conditionally properly efficient solution for (MFP).

Proof: The proof of the above theorem is similar to that of Theorem 3.2.3.

3.3 Duality in Multiobjective Fractional Programming

In relation to (MFP) we associate the following Mond-Weir type multiobjective maximization dual problem:

(MFD)   Maximize $\left(\dfrac{f_1(u)}{g_1(u)}, ..., \dfrac{f_p(u)}{g_p(u)} \right)$

subject to

$$\sum_{i=1}^{p} \tau(\nabla f_i(u) - v_i \nabla g_i(u)) + \sum_{j=1}^{m} \lambda_j \nabla h_j(u) = 0 , \quad (3.10)$$

$$\sum_{j=1}^{m} \lambda_j h_j(u) \geq 0, \quad (3.11)$$

$u \in X$, $\tau_i > 0$, $v_i \geq 0$, $i = 1,..., p$, $\lambda_j \geq 0$, $j = 1,..., m$.

Let W denote the set of all feasible solutions of the dual problem (D) and let $Y = \{u : (u, \tau, \lambda, v) \in W\}$.

We now establish weak duality and strong duality results between the primal problem (MFP) and its dual (MFD).

Theorem 3.3.1: Let x be feasible for (MFP) and (u, τ, λ, v) be feasible for (MFD). If $\left(\tau_1 \left(\dfrac{f_1}{g_1} \right), ..., \tau_p \left(\dfrac{f_p}{g_p} \right) \right)$ is $V -$ invex and $(\lambda_1 h_1, ..., \lambda_m h_m)$ is $V -$ invex with respect to the same η , then the following can not hold:

$$\frac{f_i(x)}{g_i(x)} \leq \frac{f_i(u)}{g_i(u)}, \quad \forall\ i=1,...,p,$$

$$\frac{f_{i_0}(x)}{g_{i_0}(x)} \leq \frac{f_{i_0}(u)}{g_{i_0}(u)}, \quad \text{for some } i_0 \in \{1,...,p\}.$$

Proof: Since x is feasible for (MFP) and (x,τ,λ,v) is feasible for (MFD)

$$\lambda_j h_j(x) \leq 0 \leq \lambda_j h_j(u), \quad \forall\ j=1,...,m. \tag{3.12}$$

V − invexity of $(\lambda_1 h_1,...,\lambda_m h_m)$ implies that

$$\sum_{j=1}^m \lambda_j \beta_j(x,u)\nabla h_j(u)\eta(x,u) \leq 0.$$

Since $\beta_j(x,u) > 0$, $j=1,...,m$, we get

$$\sum_{j=1}^m \lambda_j \nabla h_j(u)\eta(x,u) \leq 0.$$

From (3.10), and the above inequality, we get

$$\sum_{i=1}^p \tau_i \nabla\left(\frac{f_i(u)}{g_i(u)}\right)\eta(x,u) \geq 0.$$

Since $\alpha_i(x,u) > 0$, $i=1,...,p$, we get

$$\sum_{i=1}^p \tau_i \alpha_i(x,u)\nabla\left(\frac{f_i(u)}{g_i(u)}\right)\eta(x,u) \geq 0.$$

V − invexity of $\left(\tau_1\left(\frac{f_1}{g_1}\right),...,\tau_p\left(\frac{f_p}{g_p}\right)\right)$ implies that

$$\sum_{i=1}^p \tau_i\left(\frac{f_i(x)}{g_i(x)} - \frac{f_i(u)}{g_i(u)}\right) \geq 0.$$

Thus, the following can not hold:

$$\frac{f_i(x)}{g_i(x)} \leq \frac{f_i(u)}{g_i(u)}, \quad \forall\ i=1,...,p,$$

$$\frac{f_{i_0}(x)}{g_{i_0}(x)} \leq \frac{f_{i_0}(u)}{g_{i_0}(u)}, \quad \text{for some } i_0 \in \{1,...,p\}.$$

Theorem 3.3.2: Let x be feasible for (MFP) and (u,τ,λ,v) be feasible for (MFD). If $\left(\tau_1\left(\dfrac{f_1}{g_1}\right),...,\tau_p\left(\dfrac{f_p}{g_p}\right)\right)$ is $V-$ pseudo-invex and $(\lambda_1 h_1,...,\lambda_m h_m)$ is $V-$quasi-invex with respect to the same η, then the following can not hold:

$$\frac{f_i(x)}{g_i(x)} \le \frac{f_i(u)}{g_i(u)}, \quad \forall \; i=1,...,p,$$

$$\frac{f_{i_0}(x)}{g_{i_0}(x)} < \frac{f_{i_0}(u)}{g_{i_0}(u)}, \quad \text{for some } i_0 \in \{1,...,p\}.$$

Proof: Since x is feasible for (MFP) and (x,τ,λ,v) is feasible for (MFD) $\lambda_j h_j(x) \le 0 \le \lambda_j h_j(u), \quad \forall \; j=1,...,m.$

Since $\beta_j(x,u) > 0$, $j=1,...,m$, we get

$$\sum_{j=1}^{m} \lambda_j \beta_j(x,u) h_j(x) \le \sum_{j=1}^{m} \lambda_j \beta_j(x,u) h_j(u). \tag{3.13}$$

$V-$quasi-invexity of $(\lambda_1 h_1,...,\lambda_m h_m)$ and (3.13) implies that

$$\sum_{j=1}^{m} \lambda_j \nabla h_j(u) \eta(x,u) \le 0. \tag{3.14}$$

From (3.10), and (3.14), we get

$$\sum_{i=1}^{p} \tau_i \nabla\left(\frac{f_i(u)}{g_i(u)}\right) \eta(x,u) \ge 0. \tag{3.15}$$

$V-$pseudo-invexity of $\left(\tau_1\left(\dfrac{f_1}{g_1}\right),...,\tau_p\left(\dfrac{f_p}{g_p}\right)\right)$ and (3.15) implies that

$$\sum_{i=1}^{p} \tau_i \alpha_i(x,u)\left(\frac{f_i(x)}{g_i(x)} - \frac{f_i(u)}{g_i(u)}\right) \ge 0. \tag{3.16}$$

Since $\alpha_i(x,u) > 0$ and $\tau_i > 0$, $i=1,...,p$, therefore the following can not hold:

$$\frac{f_i(x)}{g_i(x)} \le \frac{f_i(u)}{g_i(u)}, \quad \forall \; i=1,...,p,$$

$$\frac{f_{i_0}(x)}{g_0(x)} < \frac{f_0(u)}{g_0(u)}, \quad \text{for some } i_0 \in \{1,...,p\}.$$

Theorem 3.3.3: Let x^* be feasible for (MFP) and $(u^*, \tau^*, \lambda^*, v^*)$ be feasible for (MFD) such that $v_i^* = \dfrac{f_i(x^*)}{g_i(x^*)}, \quad \forall \ i = 1,...,p..$ Let for

$$\left(\tau_1 \left(\frac{f_1}{g_1} \right),...,\tau_p \left(\frac{f_p}{g_p} \right) \right)$$

and $(\lambda_1 h_1,...,\lambda_m h_m)$ the $V-$ invexity assumption or its generalizations of Theorem 3.3.1 or Theorem 3.3.2 hold, then x^* is conditionally properly efficient solution for (MFP). Also, if for each feasible (u, τ, λ, v) for (MFD), then $(u^*, \tau^*, \lambda^*, v^*)$ is conditionally properly efficient for (MFD).

Proof: The proof of the above Theorem is similar to the proof of Theorem 2.4.3 of Chapter 2.

Theorem 3.3.4 (Strong Duality): Let x^* be a conditionally properly efficient solution for (MFP). Assume that there exists $\bar{x} \in X$ such that $h_j(x) < 0$ and $(\lambda_1 h_1,...,\lambda_m h_m)$ is $V-$ invex on X with respect to η, then there exist scalars $\tau_i^* > 0$, $i = 1,...,p$, $\lambda_j^* \geq 0$, $j = 1,...,m$ such that (x^*, τ^*, λ^*) is feasible for (MFD). Further, if for each feasible (u, τ, λ) for (MFD), $\left(\tau_1 \left(\dfrac{f_1}{g_1} \right),...,\tau_p \left(\dfrac{f_p}{g_p} \right) \right)$ is $V-$ invex at u with respect to η, then (x^*, τ^*, λ^*) is a conditionally properly efficient solution for (MFD).

Proof: Since x^* is conditionally properly efficient solution for (MFP), it follows from Theorem 3.2.1, that there exist scalars $\tau_i^* > 0$, $i = 1,...,p$, $\lambda_j^* \geq 0$, $j \in I(x^*)$ such that

$$\sum_{i=1}^{p} \tau_i^* \nabla \left(\frac{f_i(x^*)}{g_i(x^*)} \right) + \sum_{i \in I(x^*)} \lambda_i^* \nabla h_i(x^*) = 0.$$

Set $\lambda_j^* = 0$, $j \in I(x^*)$, then

$$\sum_{i=1}^{p} \tau_i^* \nabla \left(\frac{f_i(x^*)}{g_i(x^*)} \right) + \sum_{j=1}^{m} \lambda_j^* \nabla h_j(x^*) = 0 \, .$$

$$\lambda^{*T} h(x^*) = 0,$$

$$\lambda_j^* \geq 0, \quad j = 1, ..., m,$$

$$\tau_i^* \geq 0, \quad i = 1, ..., p \, .$$

Hence (x^*, τ^*, λ^*) is feasible for (MFD).

We will now prove that (x^*, τ^*, λ^*) is an efficient solution for (MFD). Suppose (x^*, τ^*, λ^*) is not an efficient solution, then there exists a feasible (u, τ, λ) of (MFD) such that

$$\frac{f_i(u)}{g_i(u)} \geq \frac{f_i(x^*)}{g_i(x^*)}, \qquad \forall \ i = 1, ..., p$$

$$\frac{f_0(u)}{g_0(u)} > \frac{f_{i_0}(x^*)}{g_0(x^*)}, \quad \text{for some } i_0 \in \{1, ..., p\}.$$

This is a contradiction to weak duality Theorem 3.3.1.

We will finally prove that (x^*, τ^*, λ^*) is a conditionally properly efficient solution for (MFD).

Suppose (x^*, τ^*, λ^*) is not a conditionally properly efficient solution for (MFD) then there exists a feasible solution (u, τ, λ) for (MFD) and an index i such that for every $M(x^*) > 0$

$$\frac{f_i(u)}{g_i(u)} > \frac{f_i(x^*)}{g_i(x^*)} \text{ and } \frac{f_i(u)}{g_i(u)} - \frac{f_i(x^*)}{g_i(x^*)} > M(x^*) \left(\frac{f_i(x^*)}{g_i(x^*)} - \frac{f_i(u)}{g_i(u)} \right)$$

such that $\dfrac{f_j(u)}{g_j(u)} < \dfrac{f_j(x^*)}{g_j(x^*)}$.

Thus $\dfrac{f_j(u)}{g_j(u)} - \dfrac{f_j(x^*)}{g_j(x^*)}$ can be made arbitrarily large and hence

$$\sum_{i=1}^{p} \tau_i^* \left(\frac{f_j(u)}{g_j(u)} - \frac{f_j(x^*)}{g_j(x^*)} \right) > 0,$$

which contradicts weak duality Theorem 3.3.1. Thus (x^*, τ^*, λ^*) is a conditionally properly efficient solution for (MFD).

Theorem 3.3.5 (Strong Duality): Let x^* be a conditionally properly efficient solution for (MFP). Assume that there exists $\bar{x} \in X$ such that $h_j(x) < 0$ and $\left(\lambda_1 h_1, ..., \lambda_m h_m\right)$ is V – pseudo-invex on X with respect to η, then there exist scalars $\tau_i^* > 0$, $i = 1, ..., p$, $\lambda_j^* \geq 0$, $j = 1, ..., m$ such that $\left(x^*, \tau^*, \lambda^*\right)$ is feasible for (MFD). Further, if for each feasible (u, τ, λ) for (MFD), $\left(\tau_1\left(\dfrac{f_1}{g_1}\right), ..., \tau_p\left(\dfrac{f_p}{g_p}\right)\right)$ is V – pseudo-invex and $\left(\lambda_1 h_1, ..., \lambda_m h_m\right)$ is V – quasi-invex at u with respect to η, then $\left(x^*, \tau^*, \lambda^*\right)$ is a conditionally properly efficient solution for (MFD).

Proof: The proof follows on the lines of the proof of Theorem 3.3.4.

We now consider the following Jagannathan type dual to multiobjective fractional programming dual problem:

(MJD)   Maximize $\left(v_1, ..., v_p\right)$

subject to

$$\sum_{i=1}^{p} \tau\left(\nabla f_i(u) - v_i \nabla g_i(u)\right) + \sum_{j=1}^{m} \lambda_j \nabla h_j(u) = 0, \quad (3.17)$$

$$\sum_{i=1}^{p} \tau\left(\nabla f_i(u) - v_i \nabla g_i(u)\right) \geq 0, \quad (3.18)$$

$$\sum_{j=1}^{m} \lambda_j h_j(u) \geq 0, \quad (3.19)$$

where $\tau, v \in R^p$, $\lambda \in R^m$. Denote $v = \left(v_1, ..., v_p\right)$ and

$$F(x) = \left(\dfrac{f_1(x)}{g_1(x)}, ..., \dfrac{f_p(x)}{g_p(x)}\right).$$

Theorem 3.3.6 (Weak Duality): Let x be feasible for (MFP) and (u, τ, λ, v) be feasible for (MJD). If $\left(f_i - v_i g_i\right)$, $i = 1, ..., p$ and $\left(\lambda_1 h_1, ..., \lambda_m h_m\right)$ are V – invex with respect to the same η, then $F(x) \nleq v$.

Proof: Suppose to the contrary that there exist x feasible for (MFP) and (u, τ, λ, v) feasible for (MFD) such that $F(x) \nleq v$. Then

$$\frac{f_i(x)}{g_i(x)} \leq v_i, \quad \forall\ i=1,...,p$$

and

$$\frac{f_{i_0}(x)}{g_{i_0}(x)} < v_{i_0}, \quad \text{for some } i_0 \in \{1,...,p\}.$$

That is,

$$f_i(x) - v_i\, g_i(x) \leq 0, \quad \forall\ i=1,...,p$$

and

$$f_{i_0}(x) - v_{i_0} g_0(x) < 0, \quad \text{for some } i_0 \in \{1,...,p\}.$$

Therefore,

$$\sum_{i=1}^{m} \tau_i \big(f_i(x) - v_i\, g_i(x)\big) < 0.$$

Using the duality constraint (3.18), we get

$$\sum_{i=1}^{m} \tau_i \big(f_i(x) - v_i\, g_i(x)\big) \leq \sum_{i=1}^{m} \tau_i \big(f_i(u) - v_i\, g_i(u)\big).$$

Using V – invexity hypothesis, we get

$$\sum_{i=1}^{m} \tau_i \alpha_i(x, u)\big(\nabla f_i(x) - v_i\, \nabla g_i(x)\big)\eta(x, u) \leq 0. \tag{3.20}$$

Now from (3.6), (3.19) and V – invexity of $\big(\lambda_i h_i,\ i \in I(x^*)\big)$, we get

$$\sum_{i \in I(x^*)} \lambda_i \beta_i(x, u)\nabla h_i(u)\eta(x, u) < 0. \tag{3.21}$$

Now, from (3.20) and (3.21), we reached to a contradiction of (3.17). Hence, $F(x) \nleq v$.

Remark 3.3.1: The above theorem holds under generalized V – invexity assumptions used in Theorem 2.4.2.

3.4 Generalized Fractional Programming

Duality results for minimax fractional programming involving several ratios in the objective function have been obtained by Crouzeix (1981), Crouzeix, Ferland and Schaible (1983, 1985), Jagannathan and Schaible (1983), Chandra, Craven and Mond (1986), Bector, Chandra and Bector

(1989), Singh and Rueda (1990), Xu (1988) and Chandra and Kumar (1993).

Crouzeix, Ferland and Schaible (1985) have shown that the minimax fractional program can be solved by solving a minimax nonlinear parametric program. Bector, Chandra and Bector (1989) have developed duality for the generalized minimax fractional program, under generalized convexity assumptions, using a minimax parametric program (see, Crouzeix, ferland and Schaible (1985)).

Recently, Bector, Chandra and Kumar (1994) have extended minimax programs under V − invexity assumptions. The purpose of this section is to extend minimax fractional programs under V − invexity assumptions and its generalizations.

Consider the following minimax fractional programming problem as the primal problem:

(P) $$v^* = \min_{x \in S} \max_{1 \le i \le p} \left[\frac{f_i(x)}{g_i(x)} \right]$$

where

(A1) $S = \{ x \in R^n : h_k(x) \le 0, \ k = 1, ..., m \}$ is nonempty and compact;

(A2) f_i, g_i, $i = 1, ..., p$ and h_k, $k = 1, ..., m$ are differentiable on R^n;

(A3) $g_i(x) > 0$, $i = 1, ..., p$, $x \in S$;

(A4) if g_i is not affine, then $f_i(x) \ge 0$ for all i and all $x \in S$.

Crouzeix, ferland and Schaible (1985) considered the following minimax nonlinear parametric programming problem in the parameter v:

(P)$_v$ $$F(v) = \min_{x \in S} \max_{1 \le i \le p} [f_i(x) - v g_i(x)].$$

The following Lemma will be needed in the sequel:

Lemma 3.4.1 (Crouzeix, Ferland and Schaible (1985)): If (P) has an optimal solution x^* with optimal value of the primal problem (P) as v^*, then $F(v^*) = 0$. Conversely, if $F(v^*) = 0$, then (P) and (P)$_{v^*}$ have the same optimal solution set.

Remark 3.4.1: In case of an arbitrary set $S \subset R^n$, Crouzeix, Ferland and Schaible (1985) showed that the optimal set of (P)$_{v^*}$ may be nonempty. In (A1), however, we have assumed $S \subset R^n$ to be compact in addition to being nonempty.

To establish the optimality and duality, we shall make use of problem $(P)_v$. We now have the following programming problem that is equivalent to $(P)_v$ for a given v:

$(EP)_v$ Minimize q

subject to

$$f_i(x) - vg_i(x) \le q, \quad i = 1, ..., p, \tag{3.22}$$

$$h_k(x) \le 0, \quad k = 1, ..., m. \tag{3.23}$$

Lemma 3.4.2: If (x, v, q) is $(EP)_v$-feasible, then x is feasible for (P). If x is feasible for (P), then there exist v and q such that (x, v, q) is feasible for $(EP)_v$.

Lemma 3.4.3: x^* is optimal for (P) with corresponding optimal value of the objective function equal to v^* if and only if (x^*, v^*, q^*) is optimal for $(EP)_v$ with corresponding optimal value of the objective function equal to zero, that is, q^*.

Theorem 3.4.1 (Necessary Optimality Conditions):
Let x^* be an optimal solution for (P) with optimal value as v^*. Let an appropriate constraint qualification hold for $(EP)_{v^*}$; see (Mangasarian (1969), Craven (1978) and Kuhn and Tucker (1951). Then, there exist $q^* \in R$, $\tau^* \in R^p$, $\lambda^* \in R^m$ such that $(x^*, v^*, \tau^*, \lambda^*)$ satisfies:

$$\sum_{i=1}^{p} \tau_i^* \left(\nabla f_i(x^*) - v^* \nabla g_i(x^*) \right) + \sum_{k=1}^{m} \lambda_k^* h_k(x^*) = 0, \tag{3.24}$$

$$\sum_{i=1}^{p} \tau_i^* \left(\nabla f_i(x^*) - v^* \nabla g_i(x^*) \right) = 0, \quad \forall i = 1, ..., p, \tag{3.25}$$

$$\lambda_k^* h_k(x^*) = 0, \quad \forall k = 1, ..., m, \tag{3.26}$$

$$f_i(x^*) - v^* g_i(x^*) \le 0, \quad \forall i = 1, ..., p, \tag{3.27}$$

$$h_k(x^*) \le 0, \quad \forall k = 1, ..., m, \tag{3.28}$$

$$\sum_{i=1}^{p} \tau_i^* = 1, \tag{3.29}$$

$$q^* = 0, \tag{3.30}$$

$$q^* \in R, \ \tau^* \in R^p, \ \lambda^* \in R^m, \ \tau^* \geq 0, \ \lambda^* \geq 0. \tag{3.31}$$

Theorem 3.4.2 (Sufficient Optimality Conditions):
Let $\left(x^*, v^*, q^*, \tau^*, \lambda^*\right)$ satisfy (3.24)-(3.31), and at x^* let

$$A = \sum_{i=1}^{p} \tau_i^* \left(f_i(x) - v^* g_i(x)\right) \tag{3.32}$$

be $V-$ pseudo-invex and

$$B = \sum_{k=1}^{m} \lambda_k^* h_k(x) \tag{3.33}$$

be $V-$ quasi-invex for all x that are feasible for $(EP)_{v^*}$. Then, x^* is optimal for (P), with corresponding optimal objective value v^*.

Proof: From (3.27), (3.28), x^* is feasible for $(EP)_{v^*}$, and from (3.28), x^* is feasible for (P). Now, all x that are feasible for $(EP)_{v^*}$ are also feasible for (P). Therefore, for x^* and any x which is feasible for $(EP)_{v^*}$, we have from (3.23), (3.31), (3.26) and since

$$\beta_k(x, x^*) > 0, \ \forall \ k = 1,...,m,$$

$$\sum_{k=1}^{m} \lambda_k^* \beta_k(x, x^*) h_k(x) \leq \sum_{k=1}^{m} \lambda_k^* \beta_k(x, x^*) h_k(x^*). \tag{3.34}$$

Using the $V-$ quasi-invexity of B, we get

$$\sum_{k=1}^{m} \lambda_k^* \nabla h_k(x^*) \eta(x, x^*) \leq 0.$$

This along with (3.24) gives

$$\sum_{i=1}^{p} \tau_i^* \left(\nabla f_i(x) - v^* \nabla g_i(x)\right) \eta(x, x^*) \geq 0. \tag{3.35}$$

Using the $V-$ pseudo-invexity of A at x^*, we get from (3.35), that for any x that is feasible for $(EP)_{v^*}$, we have

$$\sum_{i=1}^{p} \tau_i^* \alpha_i(x, x^*)(f_i(x) - v^* g_i(x)) \geq \sum_{i=1}^{p} \tau_i^* \alpha_i(x, x^*)(f_i(x^*) - v^* g_i(x^*)) \tag{3.36}$$

Using (3.22), (3.29), (3.30), (3.31) and (3.36), we get

$$q \geq 0 = q^*,$$
(3.37)

for any x and q that is feasible for $(EP)_{v^*}$.

Using (3.37) and Lemma 3.4.2, we get the result.

Theorem 3.4.3 (Sufficient Optimality Conditions): Let $\left(x^*, v^*, q^*, \tau^*, \lambda^*\right)$ satisfy (3.24)-(3.31), and at x^* let

$$A = \sum_{i=1}^{p} \tau_i^* \left(f_i(x) - v^* g_i(x)\right)$$

be V – quasi-invex and

$$B = \sum_{k=1}^{m} \lambda_k^* h_k(x)$$

be strictly V – pseudo-invex for all x that are feasible for $(EP)_{v^*}$. Then, x^* is optimal for (P), with corresponding optimal objective value v^*.

Proof: From (3.27), (3.28), x^* is feasible for $(EP)_{v^*}$, and from (3.28), x^* is feasible for (P). Now, all x that are feasible for $(EP)_{v^*}$ are also feasible for (P). Therefore, for x^* and any x which is feasible for $(EP)_{v^*}$, we have from (3.23), (3.31), (3.26) and since

$$\beta_k\left(x, x^*\right) > 0, \ \forall \ k = 1,...,m,$$

$$\sum_{k=1}^{m} \lambda_k^* \beta_k\left(x, x^*\right) h_k(x) \leq \sum_{k=1}^{m} \lambda_k^* \beta_k\left(x, x^*\right) h_k\left(x^*\right).$$

Using the strict V – pseudo-invexity of B, we get

$$\sum_{k=1}^{m} \lambda_k^* \nabla h_k\left(x^*\right) \eta\left(x, x^*\right) < 0.$$
(3.38)

From (3.38) and (3.24), we get

$$\sum_{i=1}^{p} \tau_i^* \left(\nabla f_i(x) - v^* \nabla g_i(x)\right) \eta\left(x, x^*\right) > 0.$$
(3.39)

Using the V – quasi-invexity of A at x^*, we get from (3.39), that for any x that is feasible for $(EP)_{v^*}$, we have

$$\sum_{i=1}^{p} \tau_i^* \alpha_i\left(x, x^*\right) \left(f_i(x) - v^* g_i(x)\right) \geq \sum_{i=1}^{p} \tau_i^* \alpha_i\left(x, x^*\right) \left(f_i\left(x^*\right) - v^* g_i\left(x^*\right)\right).$$
(3.40)

Comparing (3.40) with (3.36), we get the result.

3.5 Duality for Generalized Fractional Programming

On the lines of Mond and Weir (1981), for a given v, we have the following dual to $(EP)_v$:

$(DEP)_v$

$$\text{Max} \sum_{i=1}^{p} \tau_i^* \left[f_i(u) - v g_i(u) \right] \qquad (3.41)$$

subject to
$$\sum_{i=1}^{p} \tau_i \left(\nabla f_i(u) - v \nabla g_i(u) \right) + \sum_{k=1}^{m} \lambda_k h_k(u) = 0, \qquad (3.42)$$

$$\sum_{k=1}^{m} \lambda_k h_k(u) \geq 0, \qquad (3.43)$$

$$\sum_{i=1}^{p} \tau_i = 1, \qquad (3.44)$$

$$u \in R^n, \ \tau \in R^p, \ \lambda \in R^m, \ \tau \geq 0, \ \lambda \geq 0. \qquad (3.45)$$

We shall now prove duality theorems relating $(EP)_v$ with $(DEP)_v$.

Theorem 3.5.1 (Weak Duality): For a given v^*, let $(\hat{x}, \hat{q})$ be feasible for $(EP)_{v^*}$, let $(\bar{u}, \bar{\tau}, \bar{\lambda})$ be feasible for $(DEP)_v$. Let

$$A = \sum_{i=1}^{p} \bar{\tau}_i \left(f_i(\cdot) - v^* g_i(\cdot) \right)$$

be V − pseudo-invex and

$$B = \sum_{k=1}^{m} \bar{\lambda}_k h_k(\cdot)$$

be V − quasi-invex for all feasible solutions for $(EP)_{v^*}$ and $(DEP)_{v^*}$. Then $\inf(EP)_{v^*} \geq \sup(DEP)_{v^*}$.

Proof: From feasibility of $(\hat{x}, \hat{q})$ and $(\bar{u}, \bar{\tau}, \bar{\lambda})$ and since $\beta_k(\hat{x}, \bar{u}) > 0, \ \forall \ k = 1, \dots, m$, we have

$$\sum_{k=1}^{m} \bar{\lambda}_k \beta_k(\hat{x}, \bar{u}) h_k(\bar{x}) \leq \sum_{k=1}^{m} \bar{\lambda}_k \beta_k(\hat{x}, \bar{u}) h_k(\bar{u}). \qquad (3.46)$$

Using V – quasi-invexity and (3.46), we get

$$\sum_{k=1}^{m} \overline{\lambda}_k \nabla h_k (\overline{u}) \eta(\hat{x}, \overline{u}) \leq 0 .$$ (3.47)

From (3.42) and (3.47), we get

$$\sum_{i=1}^{p} \overline{\tau}_i \left(\nabla f_i (\overline{u}) - v^* \nabla g_i (\overline{u}) \right) \eta(\hat{x}, \overline{u}) \geq 0.$$ (3.48)

Using the V – pseudo-invexity of A, we get

$$\sum_{i=1}^{p} \overline{\tau}_i \alpha_i (\hat{x}, \overline{u}) \left(f_i (\hat{x}) - v^* g_i (\hat{x}) \right) \geq \sum_{i=1}^{p} \overline{\tau}_i \alpha_i \left(x, x^* \right) \left(f_i (\overline{u}) - v^* g_i (\overline{u}) \right).$$ (3.49)

Using (3.44) in conjunction with (3.22) and (3.49), we get

$$\hat{q} \geq \sum_{i=1}^{p} \overline{\tau}_i \left[f_i (\overline{u}) - v^* g_i (\overline{u}) \right],$$

that is, $\inf(EP)_{v^*} \geq \sup(DEP)_{v^*} .$

Remark 3.5.1: The above theorem can also be establish with V – quasi-invexity assumption on A and strictly V – pseudo-invexity assumption on B.

Theorem 3.5.2 (Strong Duality): Let $v^* = \min_{x \in S} \max_{1 \leq i \leq p} \left[\dfrac{f_i(x)}{g_i(x)} \right]$, and let $\left(x^*, q^* \right)$ be $(EP)_{v^*}$-optimal, at which an appropriate constraint qualification holds (see, Mangasarian (1969), Craven (1978), Kuhn and Tucker (1951)). Then, there exists $\left(\tau^*, \lambda^* \right)$ such that $\left(x^*, \tau^*, \lambda^* \right)$ is feasible for $(DEP)_{v^*}$ and the corresponding objective value of $(EP)_{v^*}$ and $(DEP)_{v^*}$ are equal. If also the hypothesis of Theorem 3.5.1 are satisfied, then $\left(x^*, q^* \right)$ and $\left(x^*, \tau^*, \lambda^* \right)$ are global optima for $(EP)_{v^*}$ and $(DEP)_{v^*}$, respectively with each objective value equal to zero.

Proof: Since $\left(x^*, q^* \right)$ is optimal for $(EP)_{v^*}$, by Theorem 3.5.1, there exists $\tau^* \in R^p$, $\lambda^* \in R^m$ such that $\left(x^*, q^*, \tau^*, \lambda^* \right)$ satisfies (3.24)-(3.31). From (3.24), (3.29), (3.31), we see that $\left(x^*, \tau^*, \lambda^* \right)$ is feasible for $(DEP)_{v^*}$. Also, we see that, from (3.25), (3.26), (3.29), (3.30), we have

$$\min q = q^* = 0 = \sum_{i=1}^{p} \tau_i^* \left(f_i(x^*) - v^* g_i(x^*) \right) \tag{3.50}$$

$$= \max \sum_{i=1}^{p} \tau_i \left(f_i(x) - v^* g_i(x) \right).$$

From Theorem 3.6.1 using (3.50) along with (3.51), we infer that (x^*, q^*) is global optimum for $(\text{EP})_{v^*}$ and (x^*, τ^*, λ^*) is global optimum for $(\text{DEP})_{v^*}$, with each objective value equal to zero.

Theorem 3.5.3 (Strict Converse Duality):

For $v^* = \min_{x \in S} \max_{1 \le i \le p} \left[\dfrac{f_i(x)}{g_i(x)} \right]$, let (x^*, q^*) be optimal for $(\text{EP})_{v^*}$ at which an appropriate constraint qualification holds (see Mangasarian (1969), Craven (1978), Kuhn and Tucker (1951)). Let $(\bar{u}, \bar{\tau}, \bar{\lambda})$ be optimal for $(\text{DEP})_{v^*}$ and V − invexity hypothesis of Theorem 3.5.1 hold. Then, $\bar{u} = x^*$; that is, (u, q^*) is $(\text{EP})_{v^*}$ -optimal with each objective value equal to zero.

Proof: If possible, let $\bar{u} \ne x^*$, we now show a contradiction. Since (x^*, q^*) is optimal for $(\text{EP})_{v^*}$, there exist $\tau^* \in R^p$, $\lambda^* \in R^m$ such that (x^*, τ^*, λ^*) is optimal for $(\text{DEP})_{v^*}$ and

$$q^* = 0 = \sum_{i=1}^{p} \tau_i^* \left(f_i(x^*) - v^* g_i(x^*) \right) = \sum_{i=1}^{p} \bar{\tau}_i \left(f_i(\bar{u}) - v^* g_i(\bar{u}) \right). \tag{3.51}$$

From feasibility condition and V − quasi-invexity of B, we reach to (3.47); and from (3.42) and (3.47), we get (3.48). Using strict V − pseudo-invexity of A, we get

$$\sum_{i=1}^{p} \tau_i^* \left(f_i(x^*) - v^* g_i(x^*) \right) > \sum_{i=1}^{p} \bar{\tau}_i \left(f_i(\bar{u}) - v^* g_i(\bar{u}) \right). \tag{3.52}$$

Using (3.44) in conjunction with (3.22) on (3.52), we obtain

$$q^* > \sum_{i=1}^{p} \bar{\tau}_i \left(f_i(\bar{u}) - v^* g_i(\bar{u}) \right), \tag{3.53}$$

which contradicts (3.51).

Following Schaible (1981) and Jagannathan (1973), we associate the following fractional (FD) and nonlinear program (D) with $(\text{DEP})_{v^*}$:

(FD)   Max $\left[\dfrac{\displaystyle\sum_{i=1}^{p} \tau_i f_i(u)}{\displaystyle\sum_{i=1}^{p} \tau_i g_i(u)} \right]$

subject to

$$\nabla \left[\dfrac{\displaystyle\sum_{i=1}^{p} \tau_i f_i(u)}{\displaystyle\sum_{i=1}^{p} \tau_i g_i(u)} + \sum_{k=1}^{m} \lambda_k h_k(u) \right] = 0,$$

$$\sum_{k=1}^{m} \lambda_k h_k(u) \geq 0,$$

$$\sum_{i=1}^{p} \tau_i = 1,$$

$$\tau \in R^p, \ \lambda \in R^m, \ \tau \geq 0, \ \lambda \geq 0.$$

(D)   Max v

subject to

$$\nabla \left[\sum_{i=1}^{p} \tau_i (f_i(u) - v g_i(u)) + \sum_{k=1}^{m} \lambda_k h_k(u) \right] = 0,$$

$$\sum_{i=1}^{p} \tau_i (f_i(u) - v g_i(u)) \geq 0,$$

$$\sum_{k=1}^{m} \lambda_k h_k(u) \geq 0,$$

$$\sum_{i=1}^{p} \tau_i = 1,$$

$$u \in R^n, \ \tau \in R^p, \ \lambda \in R^m, \ \tau \geq 0, \ \lambda \geq 0.$$

We relate (FD) and (D) with $(\text{DEP})_v$ via the following theorems, the proofs are easy and hence omitted.

Theorem 3.5.4: The following relation holds:

$$\bar{v} = \frac{\displaystyle\sum_{i=1}^{p}\bar{\tau}_i f_i(\bar{u})}{\displaystyle\sum_{i=1}^{p}\bar{\tau}_i g_i(\bar{u})} = \text{Max}\left[\frac{\displaystyle\sum_{i=1}^{p}\tau_i f_i(u)}{\displaystyle\sum_{i=1}^{p}\tau_i g_i(u)}\right], \text{ for all } (u,\tau,\lambda), \text{ feasible for}$$

(FD) if and only if

$$\text{Max}\left[\sum_{i=1}^{p}\tau_i f_i(u) - \bar{v}\sum_{i=1}^{p}\tau_i g_i(u)\right] = \sum_{i=1}^{p}\bar{\tau}_i f_i(\bar{u}) - \bar{v}\sum_{i=1}^{p}\bar{\tau}_i g_i(\bar{u}) = 0,$$

for all $(u, \bar{v}, \tau, \lambda)$ feasible for $(\text{DEP})_v$. In view of Theorem 3.5.4, we can easily verify that, for optimal v, the constraint sets of (FD) and $(\text{DEP})_v$ are equivalent.

Theorem 3.5.5: If $\left(\bar{u}, \bar{\tau}, \bar{\lambda}\right)$ is feasible for (FD) and

$$\bar{v} = \frac{\displaystyle\sum_{i=1}^{p}\bar{\tau}_i f_i(\bar{u})}{\displaystyle\sum_{i=1}^{p}\bar{\tau}_i g_i(\bar{u})}, \text{ then } \left(\bar{u}, \bar{v}, \bar{\tau}, \bar{\lambda}\right) \text{ is feasible for (D).}$$

If $\left(\bar{u}, \bar{v}, \bar{\tau}, \bar{\lambda}\right)$ is feasible for (D) and $\displaystyle\sum_{i=1}^{p}\bar{\tau}_i f_i(\bar{u}) - \bar{v}\sum_{i=1}^{p}\bar{\tau}_i g_i(\bar{u}) = 0$,

then $\left(\bar{u}, \bar{v}, \bar{\tau}, \bar{\lambda}\right)$ is feasible for (D).

Theorem 3.5.6: $\left(u^*, \tau^*, \lambda^*\right)$ is optimal for (FD), with corresponding optimal objective value v^* if and only if $\left(u^*, v^*, \tau^*, \lambda^*\right)$ is optimal for (D) with corresponding optimal objective value equal v^*. Also, at the optimal solution, we get $\displaystyle\sum_{i=1}^{p}\tau_i\left(f_i(u) - vg_i(u)\right) > 0$.

Chapter 4: Multiobjective Nonsmooth Programming

4.1 Introduction

It is well known that much of the theory of optimality in constrained optimization has evolved under traditional smoothness (differentiability) assumptions, discussed in previous chapters. As nonsmooth phenomena in optimization occur naturally and frequently, the attempts to weaken these smoothness requirements have received a great deal of attention during the last two decades (Ben-Tal and Zowe (1982), Clarke (1983), Kanniappan (1983), Jeyakumar (1987, 1991), Rockaffelar (1988), Burke (1987), Egudo and Hanson (1993), Bhatia and Jain (1994), Mishra and Mukherjee (1996). Necessary optimality conditions for nonsmooth locally Lipschitz problems have been given in terms of the Clarke generalized subdifferentials (Jeyakumar (1987), Egudo and Hanson (1993), Mishra and Mukherjee (1996)). The Clarke subdifferential method has been proved to be a powerful tool in many nonsmooth optimization problems, see for example Giorgi and others (2004).

Zhao (1992) gave some generalized invex conditions for a nonsmooth constrained optimization problems generalizing those of Hanson and Mond (1982) for differentiable problems. Following Zhao (1992), Egudo and Hanson (1993) generalized V-invexity concept of Jeyakumar and Mond (1992) to the nonsmooth setting and obtained sufficient optimality conditions for a locally Lipschitz multiobjective programming in terms of Clarke's subdifferential. Wolfe type duality results are also obtained in Egudo and Hanson (1993). Mishra and Mukherjee (1996) generalized the V-pseudo-invexity and V-quasi-invexity concepts of Jeyakumar and Mond (1992) to nonsmooth setting following Egudo and Hanson (1993).

This Chapter is organized as follows: In Section 3, we establish sufficient optimality conditions to nonsmooth context using conditional proper efficiency. Using the concept of quasi-differentials due to Borwein (1979), Fritz John and Kuhn-Tucker type sufficient optimality conditions for a feasible point to be efficient or conditionally properly efficient for a subdifferentiable multiobjective fractional problem are obtained without recourse to an equivalent V-invex program or parametric transformation. In Section 4, Mond-Weir type duality results are established for the nonsmooth multiobjective programming problem, under generalized V-invexity conditions, using conditional proper efficiency. Further, various duality results are established under similar assumptions for subdifferentiable multiobjective fractional programming problems. In Section 5, a vector valued ratio type Lagrangian is considered and vector valued saddle point results are presented under V-invexity conditions and its generalizations.

4.2 V-Invexity of a Lipshitz Function

The multiobjective nonlinear programming problem to be considered is:

(VP)   Minimize $\left(f_i(x): \ i = 1,...,p\right)$

subject to $g_j(x) \le 0, \quad j = 1,...,m,$

where $f_i : R^n \to R, \quad i = 1,...,p$ and $g_j : R^n \to R, \quad j = 1,...,m$ are locally Lipschitz functions.

The generalized directional derivative of a Lipschitz function f at x in the direction d denoted by $f^0(x; d)$ (see, e.g. Clarke (1983)) is:

$$f^0(x; d) = \limsup_{\substack{y \to x \\ \lambda \downarrow 0}} \frac{f(y + \lambda d) - f(y)}{\lambda}.$$

The Clarke generalized subgradient of f at x is denoted by $\partial f(x)$, is defined as follows: $\partial f(x) = \left\{\xi \in R^n : f^0(x; d) \ge \xi^T d \quad \text{for all } d \in R^n\right\}$.

Egudo and Hanson (1993) defined invexity for locally Lipschitz functions as follows:

Definition 4.2.1: A locally Lipschitz function $f(x)$ is said to be *invex* on $X_0 \subseteq R^n$ if for $x, u \in X_0$ there exists a function

$$\eta(x, u): X_0 \times X_0 \to R$$

such that

$$f(x) - f(u) \ge \eta(x, u)\xi, \quad \forall \ \xi \in \partial f(u).$$

Definition 4.2.2. [Zhao (1992)]: A locally Lipschitz function $f(x)$ is said to be *pseudo-invex* on $X_0 \subseteq R^n$ if for $x, u \in X_0$ there exists a function $\eta(x, u): X_0 \times X_0 \to R$ such that

$$\eta(x, u)\xi \geq 0 \Rightarrow f(x) \geq f(u), \quad \forall \ \xi \in \partial f(u).$$

Definition 4.2.3. [Zhao (1992)]: A locally Lipschitz function $f(x)$ is said to be *quasi-invex* on $X_0 \subseteq R^n$ if for $x, u \in X_0$ there exists a function $\eta(x, u): X_0 \times X_0 \to R$ such that

$$f(x) \leq f(u) \Rightarrow \eta(x, u)\xi \leq 0, \quad \forall \ \xi \in \partial f(u).$$

It is clear from the definitions that every locally Lipschitz invex function is locally Lipschitz *pseudo-invex* and locally Lipschitz quasi-invex.

Using the results of Zhao (1992), Egudo and Hanson (1993) generalized the V-invexity concept of Jeyakumar and Mond (1992) to the nonsmooth case:

Definition 4.2.4. [Egudo and Hanson (1993)]: A locally Lipschitz vector function $f: X_0 \to R^p$ is said to be $V - invex$ if there exist functions $\eta(x, u): X_0 \times X_0 \to R^n$ and

$$\alpha_i(x, u): X_0 \times X_0 \to R^+ \setminus \{0\}, \ i = 1, ..., p$$

such that for $x, u \in X_0$,

$$f_i(x) - f_i(u) \geq \alpha_i(x, u)\xi_i \, \eta(x, u), \quad \forall \ \xi \in \partial f(u), \ i = 1, ..., p.$$

Definition 4.2.5. [Mishra and Mukherjee (1996)]: A locally Lipschitz vector function $f: X_0 \to R^p$ is said to be $V - pseudo\text{-}invex$ if there exist functions $\eta(x, u): X_0 \times X_0 \to R^n$ and

$$\alpha_i(x, u): X_0 \times X_0 \to R^+ \setminus \{0\}, \ i = 1, ..., p$$

such that for $x, u \in X_0$,

$$\sum_{i=1}^{p} \xi_i \eta(x, u) \geq 0 \Rightarrow \sum_{i=1}^{p} \alpha_i(x, u) f_i(x) \geq \sum_{i=1}^{p} \alpha_i(x, u) f_i(u),$$

$$\forall \xi \in \partial f(u), i = 1, ..., p.$$

Definition 4.2.6. [Mishra and Mukherjee (1996)]: A locally Lipschitz vector function $f : X_0 \to R^p$ is said to be $V - quasi\text{-}invex$ if there exist functions $\eta(x, u) : X_0 \times X_0 \to R^n$ and

$$\alpha_i(x, u) : X_0 \times X_0 \to R^+ \setminus \{0\}, \ i = 1, ..., p$$

such that for $x, u \in X_0$,

$$\sum_{i=1}^{p} \alpha_i(x, u) f_i(x) \le \sum_{i=1}^{p} \alpha_i(x, u) f_i(u) \Rightarrow \sum_{i=1}^{p} \xi_i \eta(x, u) \le 0,$$

$$\forall \xi \in \partial f(u), i = 1, ..., p.$$

It is apparent from definitions that every $V -$ invex function is $V - pseudo\text{-}invex$ and $V - quasi\text{-}invex$.

Example 4.2.1: Consider

$$V - \text{Minimize} \quad \left(\left| \frac{2x_1 - x_2}{x_1 + x_2} \right|, \ \frac{x_1 - 2x_2}{x_1 + x_2} \right)$$

subject to $\quad x_1 - x_2 \le 0, \ 1 - x_1 \le 0, \ 1 - x_2 \le 0, \quad \alpha_i(x, u) = 1, i = 1, 2$

$\beta_i(x, u) = \dfrac{1}{3}(x_1 + x_2), j = 1, 2$ and $\eta(x, u) = \left(\dfrac{3(x_1 - 1)}{x_1 + x_2}, \dfrac{3(x_2 - 2)}{x_1 + x_2} \right)^T$.

As one can see that the generalized directional derivative of $f_1(x) = \left| \dfrac{2x_1 - x_2}{x_1 + x_2} \right|$ is:

$$f^0(x; d) = \lim_{\substack{y_1 \to x_1 \\ t \downarrow 0}} \sup t^{-1} \left[\left| \frac{2(y_1 + td) - x_2}{y_1 + td + x_2} \right| - \left| \frac{2y_1 - x_2}{y_1 + x_2} \right| \right]$$

$$= \lim_{\substack{y_1 \to x_1 \\ t \downarrow 0}} \sup t^{-1} \left[\left| \frac{3tdx_2}{(y_1 + x_2 + td)(y_1 + x_2)} \right| \right] \quad \left(\text{if} \quad \frac{2x_1 - x_2}{x_1 + x_2} \ge 0 \right)$$

$$= \frac{3dx_2}{(x_1 + x_2)^2}.$$

If we take $x_1 = 1$ and $x_2 = 2$ (i. e. for an efficient solution $(1, \ 2)$)

$$f^0(x; d) = \frac{2d}{3}.$$

If $y_2 \to x_2$, then $f^0(x; d) = -\dfrac{d}{3}$. Thus $\left(\dfrac{2d}{3}, -\dfrac{d}{3}\right) \in \partial f_1(u)$. It is

easy to see that $\left(-\dfrac{2}{9}, \dfrac{1}{9}\right) \in \partial f_2(u)$. At these particular points one can

easily see that above nonsmooth problem is $V - $ invex.
The following definitions will be needed in the sequel:

Definition 4.2.7 [Borwein (1979)]: The functional f is said to have an *upper derivative* at a point x^0 (denoted by $d^+ f(x^0; h)$) if

$$d^+ f(x^0; h) = \lim_{t \to 0^+} \frac{f(x + th) - f(x)}{t}, \text{ exists for all } h \in X.$$

Definition 4.2.8 [Borwein (1979)]: A functional f is said to be *quasi-differentiable* at x^0 if $d^+ f(x^0; h)$ exists and there is some weak* closed set $T(x^0)$ such that

$$d^+ f(x^0; h) = \max_{x^* \in T(x^0)} h^T x^*, \quad \forall \ h \in X. \tag{4.1}$$

The set $T(x^0)$ will be called quasi-differentiable.

Remark 4.2.1: If f is $V - $ invex and continuous at x^0, then (4.1) holds with $T(x^0) = \partial f(x^0)$.

The following Proposition can be established on the lines of Borwein (1979) and will be needed in the study of fractional programs.

Proposition 4.2.1: Let $\psi_1 : X \to R$ and $\psi_2 : X \to R$. If ψ_1 is $V - $ invex and non-negative at x^0 and $-\psi_2$ is $V - $ invex and positive at x^0, then $\theta(x) = \psi_1 / \psi_2$ is quasi-differentiable at x^0 with

$$T(x^0) = \frac{1}{\psi_2(x^0)} \left[\partial \psi_1(x^0) - \theta(x^0) \partial \psi_2(x^0)\right].$$

We now consider the following nondifferentiable multiobjective fractional programming problem:

(VFP)   Minimize $\left(\dfrac{f_1(x)}{g_1(x)}, ..., \dfrac{f_p(x)}{g_p(x)}\right)$

subject to

$$h_j(x) \leq 0, \quad j = 1, \ldots, m,$$ (4.2)

$$x \in X,$$ (4.3)

where $f_i, -g_i : X \to R$ are continuous and $V-$ invex and $g_i > 0$, $i = 1, \ldots, p$ and $h_j : X \to R$, $j = 1, \ldots, m$ are continuous and $V-$ invex.

4.3 Sufficiency of the Subgradient Kuhn-Tucker Conditions

In this section we show that the subgradient Kuhn-Tucker conditions are sufficient for conditionally properly efficient solutions.

Theorem 4.3.1 (Kuhn-Tucker type Sufficient Optimality Conditions): Let (u, τ, μ) satisfy the subgradient Kuhn-Tucker type necessary conditions

$$0 \in \sum_{i=1}^{p} \tau_i \partial f_i(u) + \sum_{j=1}^{m} \lambda_j \partial g_j(u),$$ (4.4)

$$\lambda_j g_j(u) = 0, \quad j = 1, \ldots, m,$$ (4.5)

$$\tau_i \geq 0, \quad \sum_{i=1}^{p} \tau_i = 1, \quad \lambda_j \geq 0.$$ (4.6)

If $(\tau_1 f_1, \ldots, \tau_p f_p)$ is $V-$ pseudo-invex and $(\lambda_1 g_1, \ldots, \lambda_m g_m)$ is $V-$ quasi-invex in nonsmooth sense, and u is feasible for (VP), then u is properly efficient for (VP).

Proof: The condition (4.4) implies that, there exist $\xi_i \in \partial f_i(u)$, $i = 1, \ldots, p$, $\varsigma_j \in g_j(u)$, $j = 1, \ldots, m$ such that $0 = \sum_{i=1}^{p} \tau_i \xi_i + \sum_{j=1}^{m} \lambda_j \varsigma_j(u)$.

Therefore,

$$0 = \sum_{i=1}^{p} \tau_i \xi_i \eta(x, u) + \sum_{j=1}^{m} \lambda_j \varsigma_j(u) \eta(x, u).$$

From (4.5) and feasibility of x, we get

$$\lambda_j g_j(x) \leq \lambda_j g_j(u), \quad j = 1, \ldots, m.$$

Since $\beta_j(x, u) > 0$, $j = 1, \ldots, m$, we get

$$\sum_{j=1}^{m} \lambda_j \beta_j(x, u) g_j(x) \leq \sum_{j=1}^{m} \lambda_j \beta_j(x, u) g_j(u).$$

Then by V – quasi-invexity of $(\lambda_1 g_1, \ldots, \lambda_m g_m)$, we get

$$\sum_{j=1}^{m} \lambda_j \varsigma_j \eta(x, u) \leq 0, \quad \forall \varsigma_j \in \partial g_j(u).$$

Thus, we have

$$\sum_{i=1}^{p} \tau_i \xi_i \eta(x, u) \geq 0, \quad \forall \xi_i \in \partial f_i(u).$$

Then by V – pseudo-invexity of $(\tau_1 f_1, \ldots, \tau_p f_p)$, we get

$$\sum_{i=1}^{p} \tau_i \alpha_i(x, u) f_i(x) \geq \sum_{i=1}^{p} \tau_i \alpha_i(x, u) f_i(u).$$

Since $\alpha_i(x, u) > 0$, $i = 1, \ldots, p$, we get

$$\sum_{i=1}^{p} \tau_i f_i(x) \geq \sum_{i=1}^{p} \tau_i f_i(u).$$

Hence by Theorem 1 of Geoffrion (1968) u is properly efficient solution for (VP).

We now state Fritz John and Kuhn-Tucker type necessary conditions (see, Bector et el. (1994)) and then we prove that these conditions are also sufficient for an efficient/ conditionally properly efficient solutions for (VFP) for V – invex functions and its generalizations.

Lemma 4.3.1 (Fritz John type Necessary Conditions): Let x^0 be an efficient solution for (VFP), then there exist $\tau = (\tau_1, \ldots, \tau_p) \in R_+^p$ and non-negative constant λ_j, $j = 1, \ldots, m$, not all zero such that

$$0 \in \sum_{i=1}^{p} \tau_i T_i(x^0) + \sum_{j=1}^{m} \lambda_j \partial h_j(x^0) + N_C(x^0), \tag{4.7}$$

$$\lambda_j h_j(x^0) = 0, \quad j = 1, \ldots, m, \tag{4.8}$$

where $T_i(x^0) = \dfrac{1}{g_i(x^0)} \left[\partial f_i(x^0) - \phi_i(x^0) \partial g_i(x^0) \right]$ and

$$\phi_i\left(x^0\right)=\frac{f_i\left(x^0\right)}{g_i\left(x^0\right)}. \tag{4.9}$$

To prove Kuhn-Tucker type necessary conditions, the following Slater's constraint qualification similar to that of Kanniappan (1983) is needed in the sequel.

For each $i=1,...,p$, suppose that there exist $x^i \in X$ such that $h_j\left(x^i\right)<0$, $j=1,...,m$ and $f_k\left(x^i\right)-\phi_k\left(x^0\right)g_k\left(x^i\right)<0$ for $k \neq i$, where x^0 is assumed to be an efficient solution of (MFP).

Lemma 4.3.2 (Kuhn-Tucker type Necessary Conditions): Let x^0 be an efficient solution for (VFP), and the above constraint qualification is met, then there exist $\tau=\left(\tau_1,...,\tau_p\right)\in R_+^p$ and λ_j, $j=1,...,m$, such that

$$0 \in \sum_{i=1}^{p}\tau_i T_i\left(x^0\right)+\sum_{j=1}^{m}\lambda_j \partial h_j\left(x^0\right)+N_C\left(x^0\right), \tag{4.10}$$

$$\lambda_j h_j\left(x^0\right)=0, \quad j=1,...,m, \tag{4.11}$$

$$\tau>0, \quad \lambda \geq 0, \tag{4.12}$$

where $T_i\left(x^0\right)=\dfrac{1}{g_i\left(x^0\right)}\left[\partial f_i\left(x^0\right)-\phi_i\left(x^0\right)\partial g_i\left(x^0\right)\right]$ and $\phi_i\left(x^0\right)=\dfrac{f_i\left(x^0\right)}{g_i\left(x^0\right)}$.

Theorem 4.3.2 (Fritz John Sufficiency): Assume that, there exists $\left(x^0,\tau^0,\lambda^0\right)$ where $\tau^0=\left(\tau_1^0,...,\tau_p^0\right)\in R_+^p$ and $\left(\lambda_1^0,...,\lambda_m^0\right)\in R^m$ such that

$$0 \in \sum_{i=1}^{p}\tau_i^0 T_i\left(x^0\right)+\sum_{j=1}^{m}\lambda_j^0 \partial h_j\left(x^0\right)+N_C\left(x^0\right), \tag{4.13}$$

$$\lambda_j^0 h_j\left(x^0\right)=0, \quad j=1,...,m, \tag{4.14}$$

$$h_j\left(x^0\right)\leq 0, \quad j=1,...,m, \tag{4.15}$$

and f_i, $-g_i$, $i=1,...,p$ and h_j, $j=1,...,m$ are $V-$ invex, for all $j \neq s$ and for $j=s$, $\lambda_s^0>0$ and h_s is strictly $V-$ invex. Then x^0 is an efficient solution for (VFP).

Proof: From (4.10), we have

$$0 \in \sum_{i=1}^{p} \tau_i T_i\left(x^0\right) + \sum_{j=1}^{m} \lambda_j \partial h_j\left(x^0\right) + N_C\left(x^0\right).$$

This implies that there exist some $\xi_i^0 \in \partial f_i\left(x^0\right)$ and $\varsigma_i^0 \in \partial g_i\left(x^0\right)$ for each $i = 1,...,p$, $\gamma_j^0 \in \partial h_j\left(x^0\right)$ for each $j = 1,...,m$, and $z^0 \in N_C\left(x^0\right)$ such that

$$0 = \sum_{i=1}^{p} \tau_i^0 \left(\frac{1}{g_i\left(x^0\right)}\right)\left[\xi_i^0 - \phi_i\left(x^0\right)\varsigma_i^0\right] + \sum_{j=1}^{m} \lambda_j^0 \gamma_j^0 + z^0. \tag{4.16}$$

This yields

$$0 = \eta\left(x, x^0\right)\left[\sum_{i=1}^{p} \tau_i^0 \left(\frac{1}{g_i\left(x^0\right)}\right)\left[\xi_i^0 - \phi_i\left(x^0\right)\varsigma_i^0\right] + \sum_{j=1}^{m} \lambda_j^0 \gamma_j^0 + z^0\right]. \tag{4.17}$$

If x^0 is not an efficient solution for (VFP), then there exists an x is feasible for (VFP) such that

$$\frac{f_i(x)}{g_i(x)} \le \frac{f_i\left(x^0\right)}{g_i\left(x^0\right)}, \quad \forall\ i = 1,...,p,$$

and

$$\frac{f_k(x)}{g_k(x)} < \frac{f_k\left(x^0\right)}{g_k\left(x^0\right)}, \quad \text{for atleast one } k.$$

That is,

$$f_i(x) - \phi_i\left(x^0\right)g_i(x) \le f_i\left(x^0\right) - \phi_i\left(x^0\right)g_i\left(x^0\right), \quad \forall\ i = 1,...,p,$$

and

$$f_k(x) - \phi_k\left(x^0\right)g_k(x) < f_k\left(x^0\right) - \phi_k\left(x^0\right)g_k\left(x^0\right), \quad \text{for atleast one } k.$$

Using V – invexity of $f_i - \phi_i\left(x^0\right)g_i$, $i = 1,...,p$, we have

$$\alpha_i\left(x, x^0\right)\left(\xi_i - \phi_i\left(x^0\right)\varsigma_i\right)\eta\left(x, x^0\right) \le 0, \quad \forall\ i = 1,...,p,$$

and

$$\alpha_k\left(x, x^0\right)\left(\xi_k - \phi_k\left(x^0\right)\varsigma_k\right)\eta\left(x, x^0\right) < 0, \quad \text{for atleast one } k.$$

Since $\alpha_i\left(x, x^0\right) > 0$, $i = 1,...,p$, we have

$$\left(\xi_i - \phi_i\left(x^0\right)\varsigma_i\right)\eta\left(x, x^0\right) \le 0, \quad \forall\ i = 1,...,p, \tag{4.18}$$

and

$$\left(\xi_k - \phi_k\left(x^0\right)\varsigma_k\right)\eta\left(x, x^0\right) < 0, \quad \text{for atleast one } k. \tag{4.19}$$

Multiplying (4.18) and (4.19) by $\tau_i^0 \left(\dfrac{1}{g_i(x^0)} \right) \geq 0, \quad i = 1,...,p$ and

then adding, we have

$$\sum_{i=1}^{p} \tau_i^0 \frac{1}{g_i(x^0)} \left(\xi_i - \phi_i(x^0)\varsigma_i \right) \eta(x, x^0) \leq 0. \tag{4.20}$$

From $\lambda_j^0 \geq 0$, $h_j(x) \leq 0$ and $\lambda_j^0 h_j(x^0) = 0$, $j = 1,...,m$, we have

$$\sum_{j=1}^{m} \lambda_j^0 h_j(x) \leq \sum_{j=1}^{m} \lambda_j^0 h_j(x^0).$$

Using $V - invexity$ hypothesis on h_j, $j = 1,...,m$, we have

$$\sum_{j=1}^{m} \beta_j(x, x^0) \lambda_j^0 \gamma_j^0 \eta(x, x^0) < 0.$$

Since $\beta_j(x, x^0) > 0$, $j = 1,...,m$, we have

$$\sum_{j=1}^{m} \beta_j(x, x^0) \lambda_j^0 \gamma_j^0 < 0. \tag{4.21}$$

Also, for $z^0 \in N_C(x^0)$, we have

$$z^0 \eta(x, x^0) \leq 0. \tag{4.22}$$

Combining (4.20), (4.21) and (4.22), we obtain

$$\eta(x, x^0) \left[\sum_{i=1}^{p} \tau_i^0 \left(\frac{1}{g_i(x^0)} \right) \left[\xi_i^0 - \phi_i(x^0)\varsigma_i^0 \right] + \sum_{j=1}^{m} \lambda_j^0 \gamma_j^0 + z^0 \right] < 0. \tag{4.23}$$

This contradicts (4.17). Hence the result follows.

Theorem 4.3.3 [Kuhn-Tucker Sufficient Optimality Conditions]:
Assume that there exists (x^0, τ^0, λ^0) where $\tau^0 = (\tau_1^0,...,\tau_p^0) \in R_+^p$ and
$(\lambda_1^0,...,\lambda_m^0) \in R^m$ such that

$$0 \in \sum_{i=1}^{p} \tau_i^0 T_i(x^0) + \sum_{j=1}^{m} \lambda_j^0 \partial h_j(x^0) + N_C(x^0), \tag{4.24}$$

$$\lambda_j^0 h_j(x^0) = 0, \quad j = 1,...,m, \tag{4.25}$$

$$h_j(x^0) \leq 0, \quad j = 1,...,m, \tag{4.26}$$

$$\tau^0 > 0, \quad \lambda^0 \geq 0 . \tag{4.27}$$

$f_i, -g_i, \ i = 1,..., p$ and $h_j, \ j = 1,..., m$ are V − invex. Then x^0 is a conditionally properly efficient solution for (VFP).

Proof. Since

$$\tau_i^0 \left(\frac{1}{g_i(x^0)} \right) > 0, \quad i = 1,..., p. \tag{4.28}$$

Therefore, in this case (4.18) and (4.19) will yield (4.20), as the following strict inequality

$$\sum_{i=1}^{p} \tau_i^0 \frac{1}{g_i(x^0)} \left(\xi_i - \phi_i(x^0) \varsigma_i \right) \eta(x, x^0) < 0 . \tag{4.29}$$

Combining (4.29), (4.21) and (4.22) we once again obtain (4.23), a contradiction to (4.17), as before.

We now suppose that x^0 is not conditionally properly efficient solution for (VFP). Therefore, for every positive function $M(\bar{x}) > 0$ there exists a feasible $\bar{x}$ for (VFP) and an index i such that

$$\frac{\dfrac{f_i(\bar{x})}{g_i(\bar{x})} - \dfrac{f_i(x^0)}{g_i(x^0)}}{\dfrac{f_j(x^0)}{g_j(x^0)} - \dfrac{f_j(\bar{x})}{g_j(\bar{x})}} > M(\bar{x}),$$

for all j satisfying $\dfrac{f_j(\bar{x})}{g_j(\bar{x})} < \dfrac{f_j(x^0)}{g_j(x^0)}, \quad \forall \ i = 1,..., p,$ whenever

$\dfrac{f_i(\bar{x})}{g_i(\bar{x})} > \dfrac{f_i(x^0)}{g_i(x^0)}$. This means $\dfrac{f_i(\bar{x})}{g_i(\bar{x})} - \dfrac{f_i(x^0)}{g_i(x^0)}$, i. e.,

$$\left(f_i(\bar{x}) - \phi_i(x^0) g_i(\bar{x}) \right) - \left(f_i(x^0) - \phi_i(x^0) g_i(x^0) \right)$$

can be made arbitrarily large and hence for $\tau > 0$ and

$$\tau_i^0 \left(\frac{1}{g_i(x^0)} \right) \geq 0, \quad i = 1,..., p ,$$

we obtain

$$\sum_{i=1}^{p} \tau_i^0 \frac{1}{g_i(x^0)} \left(\xi_i - \phi_i(x^0) \varsigma_i \right) > 0 . \tag{4.30}$$

This a contradiction to (4.29). Hence x^0 is a conditionally properly efficient solution for (VFP).

Remark 4.3.1: Fritz John and Kuhn-Tucker sufficiency can be establish under weaker V – invexity assumptions. Namely, f_i, $-g_i$, $i = 1,...,p$ are V – pseudo-invex and h_j, $j = 1,...,m$ are V – quasi-invex.

4.4 Subgradient Duality

For the problem (VP) considered in present Chapter, consider a corresponding Mond-Weir type dual problem:

(VD) V – Maximize $\left(f_i(u): i = 1,...,p\right)$

subject to

$$0 \in \sum_{i=1}^{p} \tau_i \partial f_i(u) + \sum_{j=1}^{m} \lambda_j \partial g_j(u), \qquad (4.31)$$

$$\lambda_j g_j(u) \geq 0, \quad j = 1,...,m, \qquad (4.32)$$

$$\tau_i \geq 0, \quad \sum_{i=1}^{p} \tau_i = 1, \quad \lambda_j \geq 0. \qquad (4.33)$$

Theorem 4.4.1 (Weak Duality): Let x be feasible for (VP) and let (u, τ, λ) be feasible for (VD). If $\left(\tau_1 f_1,...,\tau_p f_p\right)$ is V – pseudo-invex and $\left(\lambda_1 g_1,...,\lambda_m g_m\right)$ is V – quasi-invex in nonsmooth sense, then

$$\left(f_1(x),...,f_p(x)\right)^T - \left(f_1(u),...,f_p(u)\right)^T \notin -\text{int } R_+^p .$$

Proof: From the feasibility conditions,

$$\lambda_j g_j(x) \leq \lambda_j g_j(u), \quad j = 1,...,m.$$

Since $\beta_j(x, u) > 0$, $j = 1,...,m$, we get

$$\sum_{j=1}^{m} \lambda_j \beta_j(x, u) g_j(x) \leq \sum_{j=1}^{m} \lambda_j \beta_j(x, u) g_j(u).$$

Then, by V – quasi-invexity of $\left(\lambda_1 g_1,...,\lambda_m g_m\right)$, we get

$$\sum_{j=1}^{m} \lambda_j \eta(x, u) \varsigma_j \leq 0, \quad \forall \varsigma_j \in \partial g_j(u), \quad j = 1,...,m.$$

Since, $0 \in \sum_{i=1}^{p} \tau_i \partial f_i(u) + \sum_{j=1}^{m} \lambda_j \partial g_j(u)$, then there exist

$\xi_i \in \partial f_i(u)$, $i = 1,...,p$ and $\varsigma_j \in \partial g_j(u)$, $j = 1,...,m$, such that

$$0 = \sum_{i=1}^{p} \tau_i \xi_i + \sum_{j=1}^{m} \lambda_j \varsigma_j .$$

This implies that

$$0 = \sum_{i=1}^{p} \tau_i \xi_i \eta(x, u) + \sum_{j=1}^{m} \lambda_j \varsigma_j \eta(x, u).$$

Thus,

$$\sum_{i=1}^{p} \tau_i \xi_i \eta(x, u) \geq 0, \qquad \forall \, \xi_i \in \partial f_i(u), \quad i = 1,...,p.$$

Then, by V − pseudo-invexity of $(\tau_1 f_1,...,\tau_p f_p)$, we get

$$\sum_{i=1}^{p} \tau_i \alpha_i(x, u) f_i(x) \geq \sum_{i=1}^{p} \tau_i \alpha_i(x, u) f_i(u).$$

The conclusion now follows, since $\sum_{i=1}^{p} \tau_i = 1$ and

$$\alpha_i(x, u) > 0, \quad i = 1,...,p .$$

Theorem 4.4.2 (Strong Duality): Let x^0 be a weak minimum solution for (VP) at which a constraint qualification is satisfied. Then there exist $\tau^0 \in R^p$, $\lambda^0 \in R^m$, such that (x^0, τ^0, λ^0) is feasible for (VD). If weak duality holds between (VP) and (VD), then (x^0, τ^0, λ^0) is a weak minimum for (VD).

Proof: From the Kuhn-Tucker necessary conditions (see, e.g. Theorem 6.1.3 of Clarke (1983)), there exist $\tau \in R^p$, $\lambda \in R^m$, such that

$$0 \in \sum_{i=1}^{p} \tau_i \partial f_i(x^0) + \sum_{j=1}^{m} \lambda_j \partial g_j(x^0),$$

$$\tau_i \geq 0, \ \tau \neq 0, \ \lambda_j \geq 0, \ \lambda_j g_j(x^0) = 0, \quad j = 1,...,m .$$

Now since $\tau_i \geq 0$, $\tau \neq 0$, we can scale the τ_i, $i = 1,...,p$ and

λ_j, $j = 1,...,m$, thus $\tau_i^0 = \dfrac{\tau_i}{\displaystyle\sum_{i=1}^{p} \tau_i}$ and $\lambda_j^0 = \dfrac{\lambda_j}{\displaystyle\sum_{j=1}^{m} \lambda_j}$. Now we have

$\left(x^0, \tau^0, \lambda^0\right)$ that is feasible for (VD) such that

$$f_i(u) > f_i(x^0).$$

Since x^0 is feasible for (VP), this contradicts weak duality Theorem 4.4.1.

Theorem 4.4.3: Let $\bar{x}$ be feasible solution for (VP) and $\left(\bar{u}, \bar{\tau}, \bar{\lambda}\right)$ feasible for (VD) such that $\displaystyle\sum_{i=1}^{p} \bar{\tau}_i f_i(\bar{x}) = \sum_{i=1}^{p} \bar{\tau}_i f_i(u)$. If $\left(\bar{\tau}_1 f_1,...,\bar{\tau}_p f_p\right)$ is V − pseudo-invex and $\left(\bar{\lambda}_1 g_1,...,\bar{\lambda}_m g_m\right)$ is V − quasi-invex at $\bar{u}$, then $\bar{x}$ is properly efficient for (VP).

Proof: Let x be any feasible solution for (VP). From the weak duality theorem, $\displaystyle\sum_{i=1}^{p} \bar{\tau}_i f_i(x) \geq \sum_{i=1}^{p} \bar{\tau}_i f_i(\bar{u})$. From the assumption, we get

$\displaystyle\sum_{i=1}^{p} \bar{\tau}_i f_i(x) \geq \sum_{i=1}^{p} \bar{\tau}_i f_i(\bar{x})$. Hence by Theorem 1 in Geoffrion (1968), $\bar{x}$ is properly efficient for (VP).

Theorem 4.4.4: Let $\bar{x}$ be feasible for (VP) and $\left(\bar{u}, \bar{\tau}, \bar{\lambda}\right)$ be feasible for (VD) such that $f(\bar{x}) = f(\bar{u})$. If $\left(\bar{\tau}_1 f_1,...,\bar{\tau}_p f_p\right)$ is V − pseudo-invex and $\left(\bar{\lambda}_1 g_1,...,\bar{\lambda}_m g_m\right)$ is V − quasi-invex at $\bar{u}$ for each dual feasible $\left(\bar{u}, \bar{\tau}, \bar{\lambda}\right)$, then $\left(\bar{u}, \bar{\tau}, \bar{\lambda}\right)$ is properly efficient solution for (VD).

Proof: Assume that $\left(\bar{u}, \bar{\tau}, \bar{\lambda}\right)$ is not efficient, then there exists $\left(u^*, \tau^*, \lambda^*\right)$ feasible for (VD) such that $f_i(u^*) \geq f_i(\bar{u})$, $\forall\ i = 1,...,p$, and $f_j(u^*) \geq f_j(\bar{u})$, for some $j \in \{1,...,p\}$.

Therefore, $\displaystyle\sum_{i=1}^{p} \tau_i^* f_i(u^*) > \sum_{i=1}^{p} \tau_i^* f_i(\bar{u})$.

On using the assumption $f(\bar{x}) = f(\bar{u})$, we get

$$\sum_{i=1}^{p} \tau_i^* f_i(u^*) > \sum_{i=1}^{p} \tau_i^* f_i(\bar{x}),$$

a contradiction to weak duality theorem, since $\tau_i^* \geq 0$, $i = 1, ..., p$. Hence $(\bar{u}, \bar{\tau}, \bar{\lambda})$ is an efficient solution for (VD). Assume now that it is not properly efficient. Then there exist a feasible solution (u^*, τ^*, λ^*) and an $i \in \{1, ..., p\}$ such that $f_i(u^*) > f_i(\bar{u})$, and

$$f_i(u^*) - f_i(\bar{u}) > M(f_j(\bar{u}) - f_j(u^*)),$$

for all $M > 0$ and all $j \neq i \in \{1, ..., p\}$ satisfying $f_j(\bar{u}) > f_j(u^*)$. This means that $f_i(u^*) - f_i(\bar{u})$ can be made arbitrarily large whereas $f_j(\bar{u}) - f_j(u^*)$ is finite for all $j \neq i \in \{1, ..., p\}$. Therefore,

$$\tau_i^* (f_i(u^*) - f_i(\bar{u})) > \sum_{j \neq i}^{p} \tau_i^* (f_j(\bar{u}) - f_j(u^*)),$$

Or
$$\sum_{i=1}^{p} \tau_i^* f_i(u^*) > \sum_{i=1}^{p} \tau_i^* f_i(\bar{u}).$$

Using the assumption $f(\bar{x}) = f(\bar{u})$, we get

$$\sum_{i=1}^{p} \tau_i^* f_i(u^*) > \sum_{i=1}^{p} \tau_i^* f_i(\bar{x}).$$

This again contradicts the weak duality theorem since $\sum_{i=1}^{p} \tau_i^* = 1$.

Hence $(\bar{u}, \bar{\tau}, \bar{\lambda})$ is a properly efficient solution for (VD).

Theorem 4.4.5 (Strong Duality): Let $\bar{x}$ be a properly efficient solution for (VP) and $(\lambda_1 g_1, ..., \lambda_m g_m)$ satisfy the Kuhn-Tucker constraint qualification at $\bar{x}$. Then there exists $(\bar{\tau}, \bar{\lambda})$, such that $(\bar{u} = \bar{x}, \bar{\tau}, \bar{\lambda})$ is a feasible solution for (VD) and the objective values of (VP) and (VD) are equal. Also, if $(\tau_1 f_1, ..., \tau_p f_p)$ is V − pseudo-invex and $(\lambda_1 g_1, ..., \lambda_m g_m)$ is V − quasi-invex at u for every dual feasible solution (u, τ, λ), then (x, τ, λ) is a properly efficient solution for (VD).

Proof: Since $\bar{x}$ is an efficient solution for (VP) at which the Kuhn-Tucker condition is satisfied, there exist $\tau \in R^p$, $\lambda \in R^m$, such that

$$0 \in \sum_{i=1}^{p} \tau_i \partial f_i(\overline{x}) + \sum_{j=1}^{m} \lambda_j \partial g_j(\overline{x}),$$

$$\lambda_j \partial g_j(\overline{x}) \geq 0, \ \lambda_j \geq 0, \ j = 1, \ldots, m, \ \tau \neq 0, \ \tau_i \geq 0, i = 1, \ldots p.$$

Now since $\tau \neq 0$, $\tau_i \geq 0$, we can scale

$$\overline{\tau}_i = \frac{\tau_i}{\sum\limits_{i=1}^{p} \tau_i} \quad \text{and} \quad \overline{\lambda}_j = \frac{\lambda_j}{\sum\limits_{j=1}^{m} \lambda_j}.$$

Now we have $\left(\overline{x}, \ \overline{\tau}, \ \overline{\lambda}\right)$ that is feasible for (VD). Also, since the objective functions for both problems are the same, the values of (VP) and (VD) are equal at $\overline{x}$. Hence by Theorem 4.4.4 $\left(\overline{x}, \ \overline{\tau}, \ \overline{\lambda}\right)$ is a properly efficient solution of the dual problem (VD).

We now consider the following dual (VFD) to the primal problem (VFP).

(VFD)   Maximize $\left(t_1, \ldots, t_p\right)$

subject to

$$0 \in \sum_{i=1}^{p} \tau_i \left[\partial f_i(y) - t_i \partial g_i(y)\right] + \sum_{j=1}^{m} \lambda_j \partial h_j(y) + N_C(y) \quad (4.34)$$

$$f_i(y) - t_i g_i(y) \geq 0, \quad i = 1, \ldots, p \quad (4.35)$$

$$\lambda^T h(y) \geq 0, \quad (4.36)$$

$$\tau > 0, \ \lambda \geq 0, \ t \geq 0. \quad (4.37)$$

We denote the set of feasible solutions of (VFD) by K.

Theorem 4.4.6: (Weak Duality): Let x be feasible for (VFP) and $\left(y, \tau, \lambda, t\right)$ be feasible for (VFD) and let $\left(\tau_1 f_1, \ldots, \tau_p f_p\right)$ and $\left(-\tau_1 g_1, \ldots, -\tau_p g_p\right)$ and $\left(\lambda_1 h_1, \ldots, \lambda_m h_m\right)$ be $V-$ invex. Then $\dfrac{f(x)}{g(x)} \nleq t$.

Proof: Assume contrary to the result, i.e., $\dfrac{f(x)}{g(x)} \leq t$ and exhibit a contradiction.

Now $$\frac{f(x)}{g(x)} \leq t \Rightarrow \frac{f_i(x)}{g_i(x)} \leq t_i, \quad \forall \ i = 1, \ldots, p, \quad (4.38)$$

and $\dfrac{f_k(x)}{g_k(x)} < t_k$ for atleast one k.   (4.39)

From (4.35), (4.38) and (4.39), we have

$$f_i(x) - t_i g_i(x) \le f_i(y) - t_i g_i(y), \quad \forall \, i = 1,...,p,$$

and

$$f_k(x) - t_k g_k(x) \le f_k(y) - t_k g_k(y), \quad \text{for atleast one } \ k,$$

which along with the hypothesis of V − invexity on $(\tau_1 f_1,...,\tau_p f_p)$ and $(-\tau_1 g_1,...,-\tau_p g_p)$, we have

$$\alpha_i(x,\,y)[u_i - t_i v_i]\eta(x,\,y) \le 0, \quad \forall \, i = 1,...,p,$$ (4.40)

and

$$\alpha_k(x,\,y)[u_k - t_k v_k]\eta(x,\,y) < 0, \quad \text{for atleast one } \ k.$$ (4.41)

where for each $i = 1,...,p,$ $u_i \in \partial f_i(y)$ and $v_i \in \partial g_i(y)$. Using $\tau_i > 0, \, i = 1,...,p,$ with (4.40) and (4.41) and summing over i, we obtain

$$\sum_{i=1}^{p} \alpha_i(x,\,y)\tau_i[u_i - t_i v_i] < 0.$$

Since $\alpha_i(x,\,y) > 0, \quad \forall \, i = 1,...,p,$ therefore, we have

$$\sum_{i=1}^{p} \tau_i[u_i - t_i v_i] < 0.$$ (4.42)

The inequality $\lambda^T h(x) \le 0 \le \lambda^T h(y)$ along with $V-$ invexity on $(\lambda_1 h_1,...,\lambda_m h_m)$, we have

$$\sum_{j=1}^{m} \beta_j(x,\,y)\lambda_j w_j \eta(x,\,y) \le 0, \quad \forall \, w_j \in \partial h_j(y), \ j = 1,...,p.$$

Since $\beta_j(x,\,y) > 0, \quad \forall \ j = 1,...,p,$ we have

$$\sum_{j=1}^{m} \lambda_j w_j \eta(x,\,y) \le 0, \quad \forall \, w_j \in \partial h_j(y), \ j = 1,...,p.$$ (4.43)

Also, since $z \in N_C(y)$, therefore,

$$z^T \eta(x,\,y) \le 0.$$ (4.44)

From (4.42)-(4.44), we have

$$\left\{\sum_{i=1}^{p}\tau_i[u_i - t_i v_i] + \sum_{j=1}^{m}\lambda_j w_j + z\right\}\eta(x, y) < 0,$$

which contradicts (4.33). Hence the theorem.

Theorem 4.4.7 (Weak Duality): Let x be feasible for (VFP) and (y, τ, λ, t) be feasible for (VFD) and let $(\tau_1 f_1, ..., \tau_p f_p)$ and $(-\tau_1 g_1, ..., -\tau_p g_p)$ are $V-$ pseudo-invex and $(\lambda_1 h_1, ..., \lambda_m h_m)$ be $V-$ quasi-invex. Then $\dfrac{f(x)}{g(x)} \nleq t$.

Proof: From the feasibility conditions, we get

$$\sum_{j=1}^{m}\lambda_j h_j(x) \le \sum_{j=1}^{m}\lambda_j h(y).$$

Since $\beta_j(x, y) > 0$, $\forall \ j = 1, ..., p$, we have

$$\sum_{j=1}^{m}\lambda_j \beta_j(x, y) h_j(x) \le \sum_{j=1}^{m}\lambda_j \beta_j(x, y) h(y).$$

Then by $V-$ invexity of $(\lambda_1 h_1, ..., \lambda_m h_m)$, we have

$$\sum_{j=1}^{m}\lambda_j w_j \eta(x, y) \le 0, \quad \forall \ w_j \in \partial h_j(y), \ j = 1, ..., p. \tag{4.45}$$

Also, since $z \in N_C(y)$, therefore,

$$z^T \eta(x, y) \le 0. \tag{4.46}$$

Now, from (4.34), we have

$$0 = \sum_{i=1}^{p}\tau_i[u_i - t_i v_i] + \sum_{j=1}^{m}\lambda_j w_j + z, \tag{4.47}$$

for $u_i \in \partial f_i(y)$ and $v_i \in \partial g_i(y)$, $i = 1, ..., p$ and $w_j \in \partial h_j(y)$, $j = 1, ..., m$ and $z \in N_C(y)$.

Now from (4.45)-(4.47), we have

$$\sum_{i=1}^{p}\tau_i[u_i - t_i v_i]\eta(x, y) \ge 0. \tag{4.48}$$

By $V-$ pseudo-invexity of $(\tau_1 f_1, ..., \tau_p f_p)$ and $(-\tau_1 g_1, ..., -\tau_p g_p)$, we have

$$\sum_{i=1}^{p} \alpha_i(x,\, y)\tau_i\left[f_i(x) - t_i g_i(x)\right] \ge \sum_{i=1}^{p} \alpha_i(x,\, y)\tau_i\left[f_i(y) - t_i g_i(y)\right].$$

This yields,

$$f_i(x) - t_i g_i(x) \ge f_i(y) - t_i g_i(y), \quad \forall\ i = 1,\ldots, p,$$

and

$$f_k(x) - t_k g_k(x) > f_k(y) - t_k g_k(y), \quad \text{for atleast one}\quad k.$$

This implies, $\dfrac{f_i(x)}{g_i(x)} \ge t_i,\ \ \forall\ i = 1,\ldots, p,$ and $\dfrac{f_k(x)}{g_k(x)} > t_k$ for at least

one k. Thus, $\dfrac{f(x)}{g(x)} \nleq t$.

Corollary 4.4.1: Let x^0 be feasible for (VFP) and $\left(y^0,\, \tau^0, \lambda^0, t^0\right)$ be feasible for (VFD) such that $\dfrac{f\left(x^0\right)}{g\left(x^0\right)} = \dfrac{f\left(y^0\right)}{g\left(y^0\right)}$ and the invexity hypothesis either of Theorem 4.4.6 or of Theorem 4.4.7 hold. Then x^0 and $\left(y^0,\, \tau^0, \lambda^0, t^0\right)$ are conditionally properly efficient for (VFP) and (VFD), respectively.

Theorem 4.4.8 (Strong Duality): Let x^0 be an efficient solution for (VFP) and let the Slater type constraint qualification be met at x^0. Then there exist $\tau^0 \in R^p$, $\lambda^0 \in R^m$ and $t^0 \in R^p$ such that $\left(x^0,\, \tau^0, \lambda^0, t^0\right)$ is feasible for (VFD). If in addition, either Theorem 4.4.6 or Theorem 4.4.7 holds, then $\left(x^0,\, \tau^0, \lambda^0, t^0\right)$ is conditionally properly efficient solution for (VFD).

Proof: Since x^0 is an efficient solution for (VFP) and the Slater constraint qualification is met at x^0, therefore, by Lemma 4..3.2, there exist $\tau^0 \in R^p$, $\lambda^0 \in R^m$ such that the following hold

$$0 \in \sum_{i=1}^{p} \tau_i T_i\left(x^0\right) + \sum_{j=1}^{m} \lambda_j \partial h_j\left(x^0\right) + N_C\left(x^0\right), \tag{4.49}$$

$$\lambda_j h_j\left(x^0\right) = 0, \quad j = 1,\ldots, m, \tag{4.50}$$

$$\tau > 0, \quad \lambda \ge 0, \tag{4.51}$$

where $T_i(x^0) = \dfrac{1}{g_i(x^0)}\left[\partial f_i(x^0) - \phi_i(x^0)\partial g_i(x^0)\right]$

We now choose

$$\tau_i^0 = \frac{1}{g_i(x^0)} \text{ and } t_i = \phi_i(x^0) = \frac{f_i(x^0)}{g_i(x^0)}, \quad i = 1,...,p. \qquad (4.52)$$

Thus, we have

$$0 \in \sum_{i=1}^{p} \tau_i\left[\partial f_i(x^0) - t_i^0 \partial g_i(x^0)\right] + \sum_{j=1}^{m} \lambda_j^0 \partial h_j(x^0) + N_C(x^0),$$

$$f_i(x^0) - t_i^0 g_i(x^0) \geq 0, \quad i = 1,...,p,$$

$$\lambda_j^0 h_j(x^0) = 0, \qquad j = 1,...,m, \tau > 0, \lambda \geq 0.$$

This implies that $(x^0, \tau^0, \lambda^0, t^0)$ is feasible for (VFD). The condition (4.52) together with Corollary 4.4.1 gives that $(x^0, \tau^0, \lambda^0, t^0)$ is a conditionally properly efficient solution for (VFD).

Theorem 4.4.9 (Strict Converse Duality): Let x^0 be feasible for (VFP) and $(y^0, \tau^0, \lambda^0, t^0)$ be feasible for (VFD) with $t^0 = \dfrac{f_i(x^0)}{g_i(x^0)}$ $i = 1,...,p$ for all feasible (x, y, τ, λ, t), let $(\tau_1 f_1,...,\tau_p f_p)$ and $(-\tau_1 g_1,...,-\tau_p g_p)$ and $(\lambda_1 h_1,...,\lambda_m h_m)$ be $V-$ invex, and at least one of these be strictly $V-$ invex. Then $x^0 = y^0$.

Proof: Assume $x^0 \neq y^0$. Since $(y^0, \tau^0, \lambda^0, t^0)$ is feasible for (VFD), therefore, we have

$$\sum_{i=1}^{p} \tau_i^0\left[f_i(y^0) - t_i^0 g_i(y^0)\right] + \sum_{j=1}^{m} \lambda_j^0 h_j(y^0) \geq 0. \qquad (4.53)$$

Note that

$$\left\{\sum_{i=1}^{p} \tau_i^0\left[u_i^0 - t_i^0 v_i^0\right] + \sum_{j=1}^{m} \lambda_j^0 w_j^0 + z^0\right\}\eta(x, y) = 0, \qquad (4.54)$$

for some $u_i^0 \in \partial f_i(y^0)$ and $v_i^0 \in \partial g_i(y^0), i = 1,...,p$ and $w_j^0 \in \partial h_j(y^0)$, $j = 1,...,m$ and $z^0 \in N_C(y^0)$.

Using strict $V-$ invexity, we get

$$\sum_{i=1}^{p}\tau_i^0\left[f_i(x^0)-t_i^0 g_i(x^0)\right]-\sum_{i=1}^{p}\tau_i^0\left[f_i(y^0)-t_i^0 g_i(y^0)\right]$$

$$>\sum_{i=1}^{p}\tau_i^0\alpha_i(x^0,y^0)\left[u_i^0-t_i^0 v_i^0\right]\eta(x^0,y^0).$$

Since $\alpha_i(x^0,y^0)>0$, $i=1,...,p$, we have

$$\sum_{i=1}^{p}\tau_i^0\left[f_i(x^0)-t_i^0 g_i(x^0)\right]-\sum_{i=1}^{p}\tau_i^0\left[f_i(y^0)-t_i^0 g_i(y^0)\right] \qquad (4.55)$$

$$>\sum_{i=1}^{p}\tau_i^0\left[u_i^0-t_i^0 v_i^0\right]\eta(x^0,y^0).$$

Again, using strict V – invexity of $(\lambda_1 h_1,...,\lambda_m h_m)$, we get

$$\sum_{j=1}^{m}\lambda_j^0 h_j(x^0)-\sum_{j=1}^{m}\lambda_j^0 h_j(x^0)>\sum_{j=1}^{m}\lambda_j^0\beta_j(x^0,y^0)w_j^0\,\eta(x^0,y^0).$$

Since $\beta_j(x^0,y^0)>0$, for each $j=1,...,m$, we get

$$\sum_{j=1}^{m}\lambda_j^0 h_j(x^0)-\sum_{j=1}^{m}\lambda_j^0 h_j(x^0)>\sum_{j=1}^{m}\lambda_j^0 w_j^0\,\eta(x^0,y^0). \qquad (4.56)$$

Now adding (4.55) and (4.56), we get

$$\sum_{i=1}^{p}\tau_i^0\left[f_i(x^0)-t_i^0 g_i(x^0)\right]+\sum_{j=1}^{m}\lambda_j^0 h_j(x^0)>\sum_{i=1}^{p}\tau_i^0\left[f_i(y^0)-t_i^0 g_i(y^0)\right]$$

$$+\sum_{j=1}^{m}\lambda_j^0 h_j(y^0)+\left\{\sum_{i=1}^{p}\tau_i^0\left[u_i^0-t_i^0 v_i^0\right]+\sum_{j=1}^{m}\lambda_j^0 w_j^0\right\}\eta(x^0,y^0).$$

Using (4.54) and since $t_i^0=\dfrac{f_i(x^0)}{g_i(x^0)}$, $i=1,...,p$, $\sum_{j=1}^{m}\lambda_j^0 h_j(x^0)=0$,

the above equation yields

$$\sum_{i=1}^{p}\tau_i^0\left[f_i(y^0)-t_i^0 g_i(y^0)\right]+\sum_{j=1}^{m}\lambda_j^0 h_j(y^0)<0,$$

which contradicts (4.53). Hence the result follows.

4.5 Lagrange Multipliers and Saddle Point Analysis

Below we give, as a consequence of Theorem 4.3.1, a Lagrange multiplier theorem.

Theorem 4.5.1: If Theorem 4.3.1 holds, then equivalent multiobjective fractional programming problem (EFP) for (VFP) is given by

(EFP)   Minimize $\left(\dfrac{f_1(x) + \lambda^T h(x)}{g_1(x)}, ..., \dfrac{f_p(x) + \lambda^T h(x)}{g_p(x)} \right)$

subject to $\lambda_j h_j(x^0) = 0, \quad j = 1,...,m$

$\lambda_j \geq 0, \quad j = 1,...,m.$

Proof: Let x^0 be an efficient solution for (VFP). Then, (4.7), we have

$$0 \in \sum_{i=1}^{p} \tau_i \frac{1}{g_i(x^0)} \left[\partial f_i(x^0) - \frac{f_i(x^0)}{g_i(x^0)} \partial g_i(x^0) \right] \tag{4.57}$$

$$+ \sum_{j=1}^{m} \lambda_j \partial h_j(x^0) + N_C(x^0).$$

Also from (4.8), we have

$$\lambda_j^0 h_j(x^0) = 0, \quad j = 1,...,m. \tag{4.58}$$

Using (4.58) in (4.57) and without loss of generality, setting $\sum_{i=1}^{p} \tau_i \dfrac{1}{g_i(x^0)} = 1$ yields

$$0 \in \sum_{i=1}^{p} \tau_i \frac{1}{g_i(x^0)} [\partial f_i(x^0) + \lambda^{0^T} h(x^0) - \left(\frac{f_i(x^0) + \lambda^{0^T} h(x^0)}{g_i(x^0)} \right) \tag{4.59}$$

$$\partial g_i(x^0)] - N_C(x^0)$$

Now applying the arguments of Theorem 4.3.2 by replacing $f_i(x)$ by $f_i(x) + \lambda^T h(x)$ we get the result.

Theorem 4.5.1 suggests the vector valued Lagrangian function $L(x, \lambda)$ as $L: X \times R_+^m \to R^p$ given by

$$L(x, \lambda) = (L_1(x, \lambda),..., L_p(x, \lambda)),$$

where $L_i(x, \lambda) = \dfrac{f_i(x) + \lambda^T h(x)}{g_i(x)}, \quad i = 1,..., p.$

Definition 4.5.1: A point $(x^0, \lambda^0) \in X \times R_+^m$ is said to be a vector saddle point of the vector valued Lagrangian function $L(x, \lambda)$ if it satisfies the following conditions

$$L(x^0, \lambda) \not\geq L(x^0, \lambda^0), \qquad \forall\, \lambda \in R_+^m \tag{4.60}$$

and

$$L(x^0, \lambda^0) \not\geq L(x, \lambda^0), \qquad \forall\, x \in X. \tag{4.61}$$

Theorem 4.5.2: If (x^0, λ^0) is a vector saddle point of $L(x, \lambda)$, then x^0 is a conditionally properly efficient solution for (VFP).

Proof: Since (x^0, λ^0) is a vector saddle point of $L(x, \lambda)$, therefore, we have $L_i(x^0, \lambda) \leq L_i(x^0, \lambda^0)$,   for atleast one i and $\forall\, \lambda \in R_+^m$

$$\Rightarrow \frac{f_i(x^0) + \lambda^T h(x^0)}{g_i(x^0)} \leq \frac{f_i(x^0) + \lambda^{0T} h(x^0)}{g_i(x^0)}, \text{ for at least one } i \text{ and } \forall \lambda \in R_+^m$$

$$\Rightarrow (\lambda - \lambda^0)^T h(x^0) \leq 0, \qquad \forall\, \lambda \in R_+^m. \text{ This gives}$$

$$\lambda^{0T} h(x^0) = 0. \tag{4.62}$$

First we show that x^0 is an efficient solution for (VFP). Assume contrary, i.e., x^0 is not an efficient solution for (VFP). Therefore, there exists an $x \in X$ with $h(x) \leq 0$ and from (4.62) along with $\lambda^{0T} h(x) \leq 0$ yields

$$\frac{f_i(x) + \lambda^{0T} h(x)}{g_i(x)} \leq \frac{f_i(x^0) + \lambda^{0T} h(x^0)}{g_i(x^0)}, \quad \forall\, i = 1,...,p \text{ and } \forall\, x \in X$$

and $\dfrac{f_k(x) + \lambda^{0T} h(x)}{g_k(x)} < \dfrac{f_k(x^0) + \lambda^{0T} h(x^0)}{g_k(x^0)}$,for   at   least   one k

and $\forall \lambda \in R_+^m$.

That is, $L_i(x, \lambda^0) \leq L_i(x^0, \lambda^0)$, $\forall\, i = 1,...,p$ and $\forall\, x \in X$, and $L_k(x, \lambda^0) < L_k(x^0, \lambda^0)$, for at least one k and $\forall \lambda^0 \in R_+^m$, which is a contradiction to (4.61). Hence x^0 is an efficient solution for (VFP).

We now suppose that x^0 is not a conditionally properly efficient solution for (VFP). Therefore, there exists a feasible point x for (VFP) and an index i such that for every positive function $M(x^0) > 0$, we have

$$\frac{\dfrac{f_i(x)}{g_i(x)} - \dfrac{f_i(x^0)}{g_i(x^0)}}{\dfrac{f_j(x^0)}{g_j(x^0)} - \dfrac{f_j(x)}{g_j(x)}} > M(x^0),$$

for all j satisfying $\dfrac{f_j(x^0)}{g_j(x^0)} < \dfrac{f_j(x)}{g_j(x)}$, whenever $\dfrac{f_i(x^0)}{g_i(x^0)} > \dfrac{f_i(x)}{g_i(x)}$. This

along with (4.58) and $\lambda^{0T} h(x) \le 0$ yields

$$\frac{f_i(x) + \lambda^{0T} h(x)}{g_i(x)} < \frac{f_i(x^0) + \lambda^{0T} h(x^0)}{g_i(x^0)}, \quad \forall\, i = 1,\dots,p \text{ and } \forall\, x \in X,$$

which is a contradiction to (4.61). Hence x^0 is a conditionally properly efficient solution for (VFP).

Theorem 4.5.3: Let x^0 be a conditionally properly efficient solution for (VFP) and let at x^0 Slater type constraint qualification be satisfied. If $\left(\tau_1 f_1, \dots, \tau_p f_p\right)$ and $\left(-\tau_1 g_1, \dots, -\tau_p g_p\right)$ and $\left(\lambda_1 h_1, \dots, \lambda_m h_m\right)$ be $V-$ invex. Then there exists $\lambda \in R_+^m$ such that $\left(x^0, \lambda^0\right)$ is a vector saddle point of $L(x, \lambda)$.

Proof: Since x^0 is a conditionally properly efficient solution for (VFP), therefore, x^0 is also an efficient solution for (VFP) and since at x^0 Slater type constraint qualification is satisfied, therefore, by Lemma 4.3.2, there exist $\tau^0 \in R^p$ with $\tau^0 > 0$ and $\lambda^0 \in R_+^m$ such that the following hold:

$$0 \in \sum_{i=1}^{p} \tau_i \frac{1}{g_i(x^0)} \left[\partial f_i(x^0) - \phi_i(x^0) \partial g_i(x^0)\right] + \sum_{j=1}^{m} \lambda_j^0 \partial h_j(x^0) + N_C(x^0), \quad (4.63)$$

$$\lambda_j^0 h_j(x^0) = 0, \quad j = 1,\dots,m. \tag{4.64}$$

These yields

$$\sum_{i=1}^{p} \tau_i^0 \left(\frac{1}{g_i(x^0)}\right) \left[u_i^0 - t_i^0 v_i^0\right] + \sum_{j=1}^{m} \lambda_j^0 w_j^0 + z^0 = 0, \tag{4.65}$$

for some $u_i^0 \in \partial f_i(x^0)$ and $v_i^0 \in \partial g_i(x^0)$, $i = 1,\dots,p$ and $w_j^0 \in \partial h_j(x^0)$, $j = 1,\dots,m$ and $z^0 \in N_C(x^0)$.

Using the V − invexity assumption of the functions, we obtain

$$f_i(x^0) - \phi_i(x^0)g_i(x^0) \geq \alpha_i(x, x^0)(u_i^0 - \phi_i(x^0)v_i^0)\eta(x, x^0),$$
$$\forall i = 1, \ldots, p \text{ and } \forall x \in X$$

and $f_k(x^0) - \phi_k(x^0)g_k(x^0) > \alpha_k(x, x^0)(u_k^0 - \phi_k(x^0)v_k^0)\eta(x, x^0),$ for at least one k and $\forall x \in X$.

Since $\alpha_i(x, x^0) > 0, \quad \forall \ i = 1, \ldots, p$, we get

$$f_i(x^0) - \phi_i(x^0)g_i(x^0) \geq (u_i^0 - \phi_i(x^0)v_i^0)\eta(x, x^0), \tag{4.66}$$
$$\forall i = 1, \ldots, p \text{ and } \forall x \in X$$

and

$$f_k(x^0) - \phi_k(x^0)g_k(x^0) > (u_k^0 - \phi_k(x^0)v_k^0)\eta(x, x^0), \text{ for at least} \tag{4.67}$$
$$\text{one } k \text{ and } \forall x \in X.$$

Now for all $i = 1, \ldots, p, \ \forall \ x \in X$, we have

$$L_i(x, \lambda^0) - L_i(x^0, \lambda^0) = \frac{f_i(x) - \phi_i(x^0)g_i(x)}{g_i(x)} + \frac{\lambda^{0^T}h(x)}{g_i(x)}. \tag{4.68}$$

Multiplying (4.68) by $\tau_i, \quad i = 1, \ldots, p,$ which is chosen as

$\tau_i = \dfrac{\tau_i^0 g_i(x)}{g_i(x^0)}$ and $\sum\limits_{i=1}^{p} \tau_i^0 \left(\dfrac{1}{g_i(x^0)} \right) = 1$, we have

$$\sum_{i=1}^{p} \tau_i \left[L_i(x, \lambda^0) - L_i(x^0, \lambda^0) \right]$$

$$= \sum_{i=1}^{p} \tau_i^0 \left(\frac{1}{g_i(x^0)} \right) \left[f_i(x) - \phi_i(x^0)g_i(x) \right] + \lambda^{0^T} h(x),$$

which because of (4.65) and (4.66) gives

$$\sum_{i=1}^{p} \tau_i \left[L_i(x, \lambda^0) - L_i(x^0, \lambda^0) \right] \geq -\eta(x, x^0) \left(\sum_{j=1}^{m} \lambda_j^0 w_j^0 + z^0 \right) + \lambda^{0^T} h(x),$$

(because $h(\cdot)$ is V − invex at x^0 and $z^0 \in N_C(x^0)$).

Since $\tau \in R^P$, $\tau > 0$, therefore,

$$L_i(x^0, \lambda^0) \not\geq L_i(x, \lambda^0), \quad \forall \ x \in X.$$

The other part $L_i(x^0, \lambda) \not\geq L_i(x^0, \lambda^0), \quad \forall \ \lambda \in R_+^m$ of the vector saddle point inequality follows from

$$L_i\left(x^0, \lambda\right) - L_i\left(x^0, \lambda^0\right) = \frac{\lambda^T h\left(x^0\right)}{g_i\left(x^0\right)} \leq 0, \quad \forall\ i = 1,\ldots, p.$$

Hence $\left(x^0, \lambda^0\right)$ is a vector saddle point of $L(x, \lambda)$.

Remark 4.5.1: Theorem 4.5.3 can be established under weaker V − invexity assumptions, namely, $\left(\tau_1 f_1, \ldots, \tau_p f_p\right)$ and $\left(-\tau_1 g_1, \ldots, -\tau_p g_p\right)$ are V − pseudo-invex and $\left(\lambda_1 h_1, \ldots, \lambda_m h_m\right)$ is V − quasi-invex.

Chapter 5: Composite Multiobjective Nonsmooth Programming

5.1 Introduction

Jeyakumar and Yang (1993) considered the following convex composite multiobjective nonsmooth programming problem

(VP) $V - \text{Minimize} \left(f_1(F_1(x)), ..., f_p(F_p(x)) \right)$

subject to $x \in C$, $g_j(G_j(x)) \leq 0$, $j = 1,...,m$,

where C is a convex subset of a Banach space X, f_i, $i = 1,...,p$, g_j, $j = 1,...,m$, are real valued locally Lipschitz functions on R^n and F_i, $i = 1,...,p$, G_j, $j = 1,...,m$, are locally Lipschitz and Gateaux differentiable functions from X into R^n with Gateaux derivatives F_i', $i = 1,...,p$, G_j', $j = 1,...,m$, respectively, but are not necessarily continuous Frechet differentiable or strictly differentiable see Clarke (1983). The problem (VP) with $p = 1$ (single objective function) and continuously (Frechet) differentiability conditions has received a great deal of attention in the literature, e.g., Ioffe (1979), Ben-Tal and Zowe (1982), Burke (1987),and Fletcher (1982, 1987).

It is known that the scalar composite programming problem (see last Section of the present Chapter) provides a unified framework for studying convergence behaviour of various algorithms and Lagrangian conditions, e.g., see Burke (1985), Fletcher (1987) and Rockafellar (1988). Various first order optimality conditions of Lagrangian type were given in Jeyakumar (1991) for single objective composite problem without the continuously Frechet differentiability or the strict differentiability restrictions using an approximation scheme.

The Composite model problem (VP) is broad and flexible enough to cover many common types of multiobjective problems, see in the literature. Moreover, the model obviously includes the wide class of convex composite single objective problems, which is now recognized as fundamental for theory and computation in scalar nonsmooth optimization. To illustrate the nature of the model (VP), let us look at some examples.

Example 5.1.1. [Jeyakumar and Yang (1993)]:

Define F_i, $G_j : X^n \to R^{p+m}$, by

$$F_i(x) = (0,...,l_i(x),...0), \quad i = 1,...,p,$$
$$G_j(x) = (0,...,h_j(x),...0), \quad j = 1,...,m,$$

where $l_i(x)$, $i = 1,...,p$, and $h_j(x)$, $j = 1,...,m$, are locally Lipschitz and Gateaux differentiable functions on a Banach space X. Define f_i, $g_j : R^{p+m} \to R$, by

$$f_i(x) = x_i, \quad i = 1,...,p, \quad g_j(x) = x_{p+j}, \quad j = 1,...,m.$$

Let $C = X$. Then the composite problem (P) is the problem

(NP) $V - \text{Minimize} \ (l_1(x),...,l_p(x))$

subject to $x \in X^n, h_j(x) \le 0, \quad j = 1,...,m,$

which is a standard multiobjective differentiable nonlinear programming problem. Lagrangian optimality conditions, duality results and scalarization techniques for the standard multiobjective nonlinear programming problem have been extensively studied in the literature under convexity and generalized convexity conditions, see, e.g., Chew and Choo (1984), Rueda (1989), Komlosi (1993), Rapesak (1991), Jahn (1984, 1994) and Sawaragi, Nakayama and Tanino (1985). For fractional case see, e.g., Kaul, Suneja and Lalitha (1993), Mishra and Mukherjee (1995) and Characterizing the solution sets of pseudolinear programs, see, e.g., Jeyakumar and Yang (1994) and Mishra (1995).

The idea of this Chapter is that by studying the composite model problem (VP) a unified framework can be given for the treatment of many questions of theoretical and computational interest in multiobjective optimization. We have obtained results mainly for conditionally properly efficient solutions of the composite model problem (VP).

The outline of this Chapter is as follows: In Section 2, we present some preliminaries and obtain necessary optimality conditions of the Kuhn-Tucker type for the composite problem(VP). In Section 3, we present new sufficient optimality conditions for feasible points which satisfy Kuhn-Tucker type conditions to be efficient and conditionally properly efficient solutions of the problem (VP). These sufficient conditions are shown to hold for various classes of nonconvex programming problems. In Section 4, multiobjective duality results are presented for the problem (VP) under the assumptions of generalized convexity. In Section 5, a Lagrange multiplier theorem is established for the problem (VP), and a vector valued Lagrangian is introduced and vector valued saddle point results are also presented. In Section 6, we provide a scalarization result and various characterization of the set of conditionally properly efficient solutions for composite problems.

5.2 Necessary Optimality Conditions

A feasible point x_0 for (VP) is said to be an efficient solution (Sawaragi, Nakayama and Tanino (1985), White (1992)) if there exists no feasible x for (VP) such that $f_i(F_i(x)) \leq f_i(F_i(x_0))$, $i = 1, ..., p$ and $f_i(F_i(x)) \neq f_i(F_i(x_0))$, for some i. The feasible point x_0 is said to be a properly efficient solution (Jeyakumar and Yang (1993)) for (VP) if x_0 is efficient for (VP) and there exists a scalar $M > 0$ such that for each i,

$$\frac{f_i(F_i(x_0)) - f_i(F_i(x))}{f_j(F_j(x)) - f_j(F_j(x^0))} \leq M,$$

for some j such that $f_j(F_j(x)) > f_j(F_j(x_0))$ whenever x is feasible for (VP) and $f_i(F_i(x)) < f_i(F_i(x_0))$. The feasible point x_0 is said to be weakly efficient solution for (VP) if there exists no feasible point x for which $f_i(F_i(x_0)) > f_i(F_i(x))$, $i = 1, ..., p$. In the definition of proper efficiency the scalar M is independent of x, and it may happen that if f is unbounded such an M may not exist. Also an optimizer might be willing to trade different levels of losses for different levels of gains by different values of the decision variable x. Thus, on the lines of Singh and Hanson (1991), we extend the definition of proper efficiency to conditional proper efficiency for the composite model (VP) as follows:

The feasible point x_0 is said to be conditionally properly efficient solution for (VP) if x_0 is efficient for (VP) and there exists a positive function $M(x) > 0$ such that for each i,

$$\frac{f_i(F_i(x_0)) - f_i(F_i(x))}{f_j(F_j(x)) - f_j(F_j(x^0))} \leq M(x),$$

for some j such that $f_j(F_j(x)) > f_j(F_j(x_0))$ whenever x is feasible for (VP) and $f_i(F_i(x)) < f_i(F_i(x_0))$.

Notice that if $F : X \to R^n$ is locally Lipschitz near a point $x \in X$ and Gateaux differentiable at x and if $f : R^n \to R$ is locally Lipschitz near $F(x)$ then the continuous sublinear function defined by

$$x^{(h)} = \max\left\{ \sum_{k=1}^{n} w_k F_k'(x)h : w \in \partial f(F(x)) \right\},$$

satisfies the inequality $(f \circ F)_+'(x, h) \le \pi_x h, \quad \forall\, h \in X$.

The function π_x is called upper convex approximation of $f \circ F$ at x, (Jeyakumar and Yang (1993)).

The following necessary condition is taken from Jeyakumar and Yang (1993):

Theorem 5.2.1: For the problem (VP), assume that f_i, $i = 1,...,p$ and g_j, $j = 1,...,m$ are locally Lipschitz functions, and that F_i, $i = 1,...,p$ and G_j, $j = 1,...,m$ are locally Lipschitz and Gateaux differentiable functions. If $u \in C$ is a weakly efficient solution for (VP), then there exist Lagrange multipliers $\tau_i \ge 0$, $i = 1,...,p$ and $\lambda_j \ge 0$, $j = 1,...,m$ not all zero, satisfying

$$0 \in \sum_{i=1}^{p} \tau_i \partial f_i(F_i(u))F_i'(u) + \sum_{j=1}^{m} \lambda_j \partial g_j(G_j(u))G_j'(u) - (C - u)^+$$

and

$$\lambda_j g_j(G_j(u)) = 0, \quad j = 1,...,m.$$

The following Kuhn-Tucker type optimality conditions (KT) for (VP) are taken from Jeyakumar and Yang (1993):

$$0 \in \sum_{i=1}^{p} \tau_i \partial f_i(F_i(u))F_i'(u) + \sum_{j=1}^{m} \lambda_j \partial g_j(G_j(u))G_j'(u) - (C - u)^+$$

and

$$\tau \in R^p, \tau_i > 0, \lambda \in R^m, \lambda_j \ge 0, \lambda_j g_j(G_j(u)) = 0, \quad j = 1,...,m.$$

5.3 Sufficent Optimality Conditions for Composite Programs

In this Section, we present new conditions under which the necessary optimality conditions become sufficient for efficient and conditionally properly efficient solutions. The following null space condition is as in Jeyakumar and Yang (1993):

Let x, $u \in X$. Define $K : X \to R^{n(p+m)} := \pi R^n$ by

$$K(x) = \left(\left(F_1(x), ..., F_p(x) \right), \left(G_1(x), ..., G_p(x) \right) \right).$$

for each x, $u \in X$, the linear mapping $A_{x,u} : X \to R^{n(p+m)}$ is given by

$$A_{x,u} = (\alpha_1(x,u)F_1'(u)y, ..., \alpha_p(x,u)F_p'(u)y,$$

$$\beta_1(x,u)G_1'(u)y, ..., \beta_m(x,u)G_m'(u)y),$$

where $\alpha_i(x, u)$, $i = 1, ..., p$ and $\beta_j(x, u)$, $j = 1, ..., m$ are real positive constants.

Recall, from the generalized Farkas Lemma (Craven (1978)), that $K(x) - K(u) \in A_{x,u}(x)$ iff $A_{x,u}^T(y) = 0 \Rightarrow y^T(K(x) - K(u)) = 0$. Let us denote the null space of a function H by $N[H]$.

For each x, $u \in X$, there exist real constants $\alpha_i(x,u) > 0$, $i = 1, ..., p$ and $\beta_j(x,u) > 0$, $j = 1, ..., m$, such that

$$N[A_{x,u}] \subset N[K(x) - K(u)].$$

Equivalently, the null space condition mean that for each x, $u \in X$, there exist real constant $\alpha_i(x,u) > 0$, $\cdot i = 1, ..., p$ and $\beta_j(x,u) > 0$, $j = 1, ..., m$ and $\mu(x, u) \in X$ such that

$$F_i(x) - F_i(u) = \delta_i(x, u)F_i'(u)\mu(x, u)$$

and

$$G_j(x) - G_j(u) = \theta_j(x, u)G_j'(u)\mu(x, u).$$

For our problem, we assume the following generalized null space condition (GNC):

For each x, $u \in X$, there exist real constant $\alpha_i(x,u) > 0$, $i = 1, ..., p$ and $\beta_j(x,u) > 0$, $j = 1, ..., m$ and $\mu(x, u) \in (C - u)$ such that

$$F_i(x) - F_i(u) = \delta_i(x, u)F_i'(u)\mu(x, u)$$

and

$$G_j(x) - G_J(u) = \theta_j(x, u)G_j'(u)\mu(x, u).$$

Jeyakumar and Yang (1993) showed that the generalized null space condition is easily verified for nonconvex functions.

Theorem 5.3.1.[Mishra and Mukherjee (1995a)]: For the problem (VP), assume that f_i and g_j are V − invex functions, and F_i and G_j are locally Lipschitz and Gateaux differentiable functions. Let u be feasible for (VP). Suppose that the optimality conditions (KT) hold at u. If the generalized null space condition (GNC) hold at each feasible point x for (VP) then u is an efficient solution for (VP).

Proof: The condition

$$0 \in \sum_{i=1}^{p} \tau_i \partial f_i(F_i(u))F_i'(u) + \sum_{j=1}^{m} \lambda_j \partial g_j(G_j(u))G_j'(u) - (C-u)^+,$$

implies there exist $v_i \in \partial f_i(F_i(u)), \ i=1,...,p$ and $w_j \in \partial g_j(G_j(u)),$ $j = 1,...,m$ such that

$$\sum_{i=1}^{p} \tau_i v_i^T F_i'(u) + \sum_{j=1}^{m} \lambda_j w_j^T G_j'(u) \in (C-u)^+.$$

Suppose that u is not an efficient solution for (VP). Then there exists a feasible $x \in C$ for (VP) with $f_i(F_i(x)) \le f_i(F_i(u)), \ i=1,...,p$, and $f_{i_0}(F_{i_0}(x)) < f_{i_0}(F_{i_0}(u))$, for some $i_0 \in \{1,...,p\}$.

Now, by the generalized null space condition, there exists $\mu(x, u) \in (C-u)$, same for each F_i and G_j, such that

$$F_i(x) - F_i(u) = \delta_i(x, u)F_i'(u)\mu(x, u), \ i=1,...,p$$

and

$$G_j(x) - G_J(u) = \theta_j(x, u)G_j'(u)\mu(x, u), \ j=1,...,m$$

and by V − invexity of f_i and g_j there exists $\eta(x,u),\alpha_i(x,u) > 0$ $i=1,...,p$ and $\beta_j(x,u) > 0, j=1,...,m$ such that

$$f_i(F_i(x)) - f_i(F_i(u)) - \alpha_i(x,u)\xi_i\eta(x,u), \forall \xi_i \in \partial f_i(F_i(u)), i=1,...,p$$

and

$$g_j(G_j(x)) - g_j(G_j(u)) - \beta_j(x,u)\varsigma_j\eta(x,u),$$
$$\forall \varsigma_j \in \partial g_j(G_j(u)), j=1,...,m.$$

Hence

$$0 \geq \sum_{j=1}^{m} \frac{\lambda_j}{\beta_j(x, u)\theta_j(x, u)}\left(g_j\left(G_j(x)\right) - g_j\left(G_j(u)\right)\right) \quad \text{(by feasibility)}$$

$$\geq \sum_{j=1}^{m} \frac{\lambda_j}{\theta_j(x, u)}\varsigma_j \eta(x, u)\left(G_j(x) - G_j(u)\right), \quad \forall \ \varsigma_j \in \partial g_j\left(G_j(u)\right)$$

$$\text{(by subdifferentiability)}$$

$$= \sum_{j=1}^{m} \lambda_j \varsigma_j \eta(x, u)G_j'(u)\mu(x, u), \quad \text{(by (GNC))}$$

$$\geq -\sum_{i=1}^{p} \tau_i \xi_i \eta(x, u)F_i'(u)\mu(x, u), \quad \text{(by hypothesis)}$$

$$\geq \sum_{i=1}^{p} \frac{\tau_i}{\alpha_i(x, u)\delta_i(x, u)}\left(f_i\left(F_i(x)\right) - f_i\left(F_i(u)\right)\right)$$

$$\text{(by subdifferentiability)}$$

$$> 0.$$

This is a contradiction and u is an efficient solution for (VP).

Theorem 5.3.2: For the problem (VP), assume that $\left(\tau_1 f_1(\cdot), ..., \tau_p f_p(\cdot)\right)$ is $V-$pseudo-invex and $\left(\lambda_1 g_1(\cdot), ..., \lambda_m g_m(\cdot)\right)$ is $V-$quasi-invex and F_i and G_j are locally Lipschitz and Gateaux differentiable functions. Let u be feasible for (VP). Suppose that the optimality conditions (KT) hold at u. If the generalized null space condition (GNC) hold at each feasible point x for (VP) then u is an efficient solution for (VP).

Proof: As in the proof of above theorem, we have

$$\sum_{i=1}^{p} \alpha_i(x, u)\tau_i f_i\left(F_i(x)\right) \leq \sum_{i=1}^{p} \alpha_i(x, u)\tau_i f_i\left(F_i(u)\right).$$

Now by $V-$pseudo-invexity of $\left(\tau_1 f_1(\cdot), ..., \tau_p f_p(\cdot)\right)$ and the generalized null space condition (GNC), we get

$$\sum_{i=1}^{p} \tau_i \xi_i \eta(x, u)F_i'(u)\mu(x, u) \leq 0, \quad \forall \ \xi_i \in \partial f_i\left(F_i(u)\right),$$

with at least one strict inequality.

So by hypothesis, we have

$$\sum_{j=1}^{m} \lambda_j \varsigma_j \eta(x, u)G_j'(u)\mu(x, u) > 0, \quad \forall \ \varsigma_j \in \partial g_j\left(G_j(u)\right).$$

Then by $V-$quasi-invexity of $(\lambda_1 g_1(\cdot),\ldots,\lambda_m g_m(\cdot))$ and the generalized null space condition, we get

$$\sum_{j=1}^{m}\lambda_j g_j\left(G_j(x)\right)>\sum_{j=1}^{m}\lambda_j g_j\left(G_j(u)\right).$$

This is a contradiction, since

$$\lambda_j g_j\left(G_j(x)\right)\leq 0,\qquad \lambda_j g_j\left(G_j(u)\right)=0 \ .$$

Theorem 5.3.3: If u is an optimal solution of

$$\text{(VP}_\tau)\qquad \text{Minimize } \sum_{i=1}^{p}\tau_i f_i\left(F_i(x)\right)$$

$$\text{subject to } x\in c,\ \lambda_j g_j\left(G_j(x)\right)\leq 0, j=1,\ldots,m,$$

then u is conditionally properly efficient solution for (VP).

Proof: Obviously u is efficient. Choose a function $M(x)$ such that

$$M(x)=(p-1)\max_{i,j}\left(\frac{\tau_j(x)}{\tau_i(x)}\right).$$

Suppose u is not conditionally properly efficient. Then for some i and j, $f_i(F_i(x))-f_i(F_i(u))>M(x)(f_j(F_j(u))-f_j(F_j(x)))$.

That is,

$$f_i(F_i(x))-f_i(F_i(u))>(p-1)\max_{i,j}\left(\frac{\tau_j(x)}{\tau_i(x)}\right)(f_j(F_j(u))-f_j(F_j(x)))$$

$$>(p-1)\left(\frac{\tau_j(x)}{\tau_i(x)}\right)(f_j(F_j(u))-f_j(F_j(x))).$$

Thus,

$$\frac{\tau_i}{(p-1)}(f_i(F_i(x))-f_i(F_i(u)))>\tau_j\left(f_j(F_j(u))-f_j(F_j(x))\right).$$

Summing over $j\neq i$,

$$\tau_i(f_i(F_i(x))-f_i(F_i(u)))>\sum_{j\neq i}\tau_j\left(f_j(F_j(u))-f_j(F_j(x))\right).$$

That is,

$$\tau_i f_i(F_i(x))+\sum_{j\neq i}\tau_j f_j(F_j(x))>\tau_i f_i(F_i(u))+\sum_{j\neq i}\tau_j f_j(F_j(u)),$$

which, since $\tau_i>0,\ \ i=1,\ldots,p$, contradicts the optimality of u.

Hence u is conditionally properly efficient.

Theorem 5.3.4: Assume that the conditions on (VP) in Theorem 5.3.1 hold. Let u be feasible for (VP). Suppose that the optimality conditions (KT) hold at u. If the generalized null space condition (GNC) holds with $\delta_i(x, u) = \theta_j(x, u) = 1$, $\forall$ i, j, for each feasible x of (VP), then u is a conditionally properly efficient solution for (VP).

Proof: Let x be feasible for (VP). Then x is also feasible for the scalar problem (VP$_\tau$). From the V – invexity property of f_i, $i = 1,..., p$, we get

$$\sum_{i=1}^{p} \tau_i f_i(F_i(x)) - \sum_{i=1}^{p} \tau_i f_i(F_i(u))$$

$$\geq \sum_{i=1}^{p} \tau_i \alpha_i(x, u) \xi_i \eta(x, u)(F_i(x) - F_i(u)), \ \forall \ \xi_i \in \partial f_i(F_i(u)).$$

Now, by the Generalized null space condition, we get

$$\sum_{i=1}^{p} \tau_i f_i(F_i(x)) - \sum_{i=1}^{p} \tau_i f_i(F_i(u))$$

$$\geq \sum_{i=1}^{p} \tau_i \alpha_i(x, u) \xi_i \eta(x, u)(F_i(x) - F_i(u)), \ \forall \ \xi_i \in \partial f_i(F_i(u)).$$

$$= \sum_{i=1}^{p} \tau_i \alpha_i(x, u) \xi_i \eta(x, u) F_i'(u) \mu(x, u)$$

$$\geq -\sum_{j=1}^{m} \lambda_j \beta_j(x, u) \varsigma_j \eta(x, u) G_j'(u) \mu(x, u)$$

$$\geq -\sum_{j=1}^{m} \lambda_j g_j(G_j(x)) + \sum_{j=1}^{m} \lambda_j g_j(G_j(u))$$

$$\geq 0,$$

and so u is minimum for the scalar problem (VP$_\tau$). Since $\tau \neq 0 \in R^p$, $\tau > 0$, it follows from Theorem 5.3.3, that u is a conditionally properly efficient solution for (VP).

The following numerical example provides a nonsmooth composite problem for which our sufficiency Theorem 5.3.1 is satisfied.

Example 5.3.1: Consider the multiobjective problem

$$V - \text{Minimize} \ \left(\left| \frac{2x_1 - x_2}{x_1 + x_2} \right|, \ \frac{x_1 + 2x_2}{x_1 + x_2} \right)$$

subject to $x_1 - x_2 \leq 0, 1 - x_1 \leq 0, 1 - x_2 \leq 0$

Let $F_1(x) = \dfrac{2x_1 - x_2}{x_1 + x_2}$, $F_2(x) = \dfrac{x_1 + 2x_2}{x_1 + x_2}$, $G_1(x) = x_1 - x_2$, $G_2(x) = 1 - x_1$,

$G_3(x) = 1 - x_2$, $f_1(y) = |y|$, $f_2(y) = y$, $g_1(y) = g_2(y) = g_3(y) = y$,

$\alpha_i(x, u) = 1$, $i = 1, 2$, $\beta_j(x, u) = \dfrac{1}{3}(x_1 + x_2)$, $j = 1, 2, 3$ and

$$\eta(x, u) = \left(\frac{3(x_1 - 1)}{x_1 + x_2}, \frac{3(x_2 - 2)}{x_1 + x_2} \right).$$

Then the problem becomes a nonconvex composite problem with an efficient solution $(1, 2)$. It is easy to see that the null space condition holds at each feasible point of the problem. The optimality conditions (KT) also hold with

$$\xi_i = 1, \ i = 1, 2, \ \tau_1 = 1, \ \tau_2 = 3, \varsigma_j = 1, \ \lambda_j = 0, \ j = 1, 2, 3.$$

We shall now give some classes of nonlinear problems which satisfy our sufficient conditions.

Example 5.3.2. (η – **Pseudolinear programming problem**): Consider the multiobjective η – pseudolinear programming problem

(GPLP) V – Minimize $(l_1(x), ..., l_p(x))$

subject to $x \in R^n$, $h_j(x) - b_j \leq 0$, $j = 1, ..., m$,

where $l_i : R^n \to R$ and $h_j : R^n \to R$ are differentiable and η – pseudolinear i.e., pseudo-invex and pseudo-incave (Kaul, Suneja and Lalitha (1993)), and $b_j \in R$, $j = 1, ..., m$. It should be noted that a real-valued function $h : R^n \to R$ is η – pseudolinear if and only if for each $x, u \in R^n$, there exists a real constant $\alpha(x, u) > 0$ and $\eta : R^n \times R^n \to R$ such that

$$h(u) = h(x) + \alpha(x, u)h'(x)\eta(x, u).$$

For further details about pseudolinear and η – pseudolinear functions and programs see, e.g., Chew and Choo (1984), Rueda (1989), Rapesak (1991), Komlosi (1993), Kaul, Suneja and Lalitha (1993), Mishra and Mukherjee (1996b), Mishra (1995c) and Mishra, Wang and Lai (2006-2007).

Define $F_i, G_j : R^n \to R^{p+m}$ by $F_i(x) = (0, 0, ..., l_i(x), 0, ..., 0)$, $i = 1, ..., p$ and $G_j(x) = (0, 0, ..., h_j(x) - b_j, 0, ..., 0)$, $j = 1, ..., m$. Define $f_i, g_j : R^{p+m} \to R$ by $f_i(x) = x_i$, $i = 1, ..., p$, $g_j(x) = x_{p+j}$, $j = 1, ..., m$, Then, we can rewrite (GPLP) as the following nonconvex composite multiobjective problem:

$$V - \text{Minimize } (f_1(F_1(x)), ..., f_p(F_p(x)))$$

$$\text{subject to } x \in R^n, \ g_j(G_j(x)) \le 0, \ j = 1, ..., m.$$

Now, our generalized null space condition is verified at each feasible point by the η – pseudolinearity property of the functions involved. It follows from our sufficiency results that if the optimality conditions

$$\sum_{i=1}^{p} \tau_i l_i'(u) + \sum_{j=1}^{m} \lambda_j g_j'(u) = 0, \quad \lambda_j (g_j(u) - b_j) = 0, \quad \text{hold with} \quad \tau_i > 0,$$

$i = 1, ..., p$ and $\lambda_j \ge 0$, $j = 1, ..., m$ at the feasible point $u \in R^n$ of (GPLP) then u is an efficient solution for (GPLP).

We now see that the sufficient optimality conditions given in Theorem 5.3.1 holds for a class of nonconvex composite η – pseudolinear programming problem.

Example 5.3.3: Consider the problem

$$V - \text{Minimize } (f_1((h \circ \psi)(x)), ..., f_p((h \circ \psi)(x)))$$

$$\text{subject to } x \in X, \ g_j((h \circ \psi)(x)) \le 0, \ j = 1, ..., m,$$

where $h = (h_1, ..., h_n)$ is a η – pseudolinear vector function from X to R^n, ψ is a Frechet differentiable mapping from X onto itself such that $\psi'(u)$ is surjective for each $u \in X$, and f_i, g_j are V – invex. For this class of nonconvex problems, the generalized null space condition holds. To see this, let $x, u \in R^n$, $y = \psi(x)$ and $z = \psi(u)$. Then, by the η – pseudolinearity, we get

$$h_i(\psi(x)) - h_i(\psi(u)) = h_i(y) - h_i(z) = \alpha_i(y, z) h_i'(z) \eta(y, z).$$

Since $\psi'(u)$ is onto, $\eta(y, z) = \psi'(u) \xi(x, u)$ is solvable for some $G(x, u) \in R^n$. Hence,

$$h_i\big(\psi(x)\big)-h_i\big(\psi(u)\big)=\alpha_i\big(y,z\big)h_i'\big(z\big)\psi_i'\big(w\big)G(x,u)$$

$$=\hat{\alpha}_i\big(x,u\big)\big(h_i\circ\psi\big)'\big(U\big)G(x,u)$$

where $\hat{\alpha}_i(x,u)=\alpha_i\big(\psi(x),\psi(u)\big)>0$; thus (GNC) holds.

We finish this Section by observing that any finite dimensional nonconvex programming problem can also be rephrased as a composite problem (VP) and it clearly satisfies the generalized null space condition.

5.4 Subgradient Duality for Composite Multiobjective Programs

For the composite multiobjective programming problem (VP) considered in Section 5.1 above, we have the following Mond-Weir type dual:

(VD) 　$V-\text{Maximize}\ \big(f_1\big(F_1(u)\big),...,f_p\big(F_p(u)\big)\big)$

subject to

$$0\in\sum_{i=1}^{p}\tau_i\partial f_i\big(F_i(u)\big)F_i'(u)+\sum_{j=1}^{m}\lambda_j\partial g_j\big(G_j(u)\big)G_j'(u)-(C-u)^+$$

$$\lambda_j g_j\big(G_j(u)\big)\ge 0,\quad j=1,\,...,m,$$

$$u\in C,\ \tau\in R^p,\ \tau_i>0,\ \lambda\in R^m,\ \lambda_j\ge 0.$$

The following Theorems 5.4.1-5.4.5 are from Mishra and Mukherjee (1995):

Theorem 5.4.1 (Weak Duality): Let x be feasible for (VP) and let (u,τ,λ) be feasible for (VP). Assume that the generalized null space condition (GNC) holds. If $(f_1,...,f_p)$ and $(g_1,...,g_m)$ are $V-$invex and F_i, $i=1,...,p$ and G_j, $j=1,...,m$ are locally Lipschitz and Gateaux differentiable functions. Then,

$$\big(f_1\big(F_1(x)\big),...,f_p\big(F_p(x)\big)\big)^T-\big(f_1\big(F_1(u)\big),...,f_p\big(F_p(u)\big)\big)^T\notin-R_+^p\setminus\{0\}.$$

Proof: Since (u,τ,λ) is feasible for (VD), there exist

$$\tau>0,\ \lambda\ge 0,\ v_i\in\partial f_i\big(F_i(u)\big),\ i=1,...,p,$$

$$w_j\in\partial g_j\big(G_j(u)\big),\ j=1,...,m,$$

satisfying $\lambda_j g_j\big(G_j(u)\big)\ge 0,\quad j=1,...,m,$ and

$$\sum_{i=1}^{p} \tau_i v_i^T F_i'(u) + \sum_{j=1}^{m} \lambda_j w_j^T G_j'(u) \in (C-u)^+ .$$

Suppose that $x \neq u$ and

$$\left(f_1(F_1(x)),...,f_p(F_p(x))\right)^T - \left(f_1(F_1(u)),...,f_p(F_p(u))\right)^T \in -R_+^p \setminus \{0\}.$$

Then $\displaystyle\sum_{i=1}^{p} \frac{\tau_i}{\alpha_i(x,u)\delta_i(x,u)} \left(f_i(F_i(x)) - f_i(F_i(u))\right) < 0$,

Since $\displaystyle\frac{\tau_i}{\alpha_i(x,u)\delta_i(x,u)} > 0$.

Now, by the $V-$invexity of f_i and by the generalized null space condition (GNC), we get $\displaystyle\sum_{i=1}^{p} \tau_i \xi_i \eta(x,u) F_i'(u)\mu(x,u) < 0.$

From the feasibility conditions, we get $\lambda_j g_j(G_j(x)) \leq \lambda_j g_j(G_j(u))$,

and so $\displaystyle\sum_{j=1}^{m} \frac{\lambda_j}{\beta_j(x,u)\theta_j(x,u)} \left(g_j(G_j(x)) - g_j(G_j(u))\right) \leq 0.$

By $V-$invexity of g_j, $\beta_j(x,u) > 0$, $\theta_j(x,u) > 0$ and the generalized null space condition (GNC), we get

$$\sum_{j=1}^{m} \lambda_j \varsigma_j \eta(x,u) G_j'(u)\mu(x,u) \leq 0 , \quad \forall \; \varsigma_j \in \partial g_j(G_j(u)).$$

Hence $\displaystyle\left[\sum_{i=1}^{p} \tau_i \xi_i F_i'(u) + \sum_{j=1}^{m} \lambda_j \varsigma_j G_j'(u)\right]\mu(x,u)\eta(x,u) < 0.$

This is a contradiction. The proof is complete by noticing that when $x = u$ the conclusion trivially holds.

Theorem 5.4.2 (Weak Duality): Let x be feasible for (VP) and let (u, τ, λ) be feasible for (VP). Assume that the generalized null space condition (GNC) holds. If $\left(\tau_1 f_1,...,\tau_p f_p\right)$ is $V-$pseudo-invex and $\left(\lambda_1 g_1,...,\lambda_m g_m\right)$ is $V-$quasi-invex and F_i, $i = 1,...,p$ and G_j, $j = 1,...,m$ are locally Lipschitz and Gateaux differentiable functions. Then,

$$\left(f_1(F_1(x)),...,f_p(F_p(x))\right)^T - \left(f_1(F_1(u)),...,f_p(F_p(u))\right)^T \notin -R_+^p \setminus \{0\}.$$

Proof: From the feasibility conditions, we get

$$\sum_{j=1}^{m} \frac{\lambda_j}{\beta_j(x, u)\theta_j(x, u)} \big(g_j\big(G_j(x)\big) - g_j\big(G_j(u)\big)\big) \le 0.$$

Then by V – quasi-invexity of $(\lambda_1 g_1, ..., \lambda_m g_m)$ and the generalized null space condition, we have

$$\sum_{j=1}^{m} \lambda_j \varsigma_j \eta(x, u) G_j'(u)\mu(x, u) \le 0, \quad \forall \; \varsigma_j \in \partial g_j\big(G_j(u)\big).$$

Hence by the hypothesis, we have $\displaystyle\sum_{i=1}^{p} \tau_i \xi_i \eta(x, u) F_i'(u)\mu(x, u) \ge 0.$

The conclusion now follows from the V – pseudo-invexity of $(\tau_1 f_1, ..., \tau_p f_p)$.

The following two theorems can be proved as Theorem 2 and Theorem 3 of Singh and Hanson (1991).

Theorem 5.4.3: If u is optimal for (VP$_\tau$), then there exists ν such that (u, ν) is optimal for (VD$_\tau$).

Theorem 5.4.4: If u is optimal for (VP$_\tau$), then u is conditionally properly efficient for (VP), and there exists ν such that (u, ν) is conditionally properly efficient for (VD).

Theorem 5.4.5 (Strong Duality): For the problem (VP), assume that the generalized Slater constraint qualification in Section 2 holds and that the generalized null space condition (GNC) is verified at each feasible point of (VP) and (VD). If u is conditionally properly efficient solution for (VP), then there exists $\tau \in R^p$, $\tau_i > 0$, $\lambda \in R^m$, $\lambda_j \ge 0$ such that (u, τ, λ) is a conditionally properly efficient solution for (VD) and the objective values at these points are equal.

Proof: It follows from Theorem 5.2.1 that there exist $\tau \in R^p, \tau_i > 0$, $\lambda \in R^m, \lambda_j \ge 0$, such that

$$0 \in \sum_{i=1}^{p} \tau_i \partial f_i\big(F_i(u)\big) F_i'(u) + \sum_{j=1}^{m} \lambda_j \partial g_j\big(G_j(u)\big) G_j'(u) - (C - u)^+$$
$$\lambda_j g_j\big(G_j(u)\big) \ge 0, \quad j = 1, ..., m.$$

Then (u, τ, λ) is a feasible solution for (VD). From the weak duality theorem, the point (u, τ, λ) is an efficient solution for (VD).

We shall now prove that (u, τ, λ) is a conditionally properly efficient solution for (VD). Suppose that (u, τ, λ) is not conditionally properly efficient solution for (VD). Then there exists (u^*, τ^*, λ^*) feasible for (VD) such that

$$f_i\left(F_i\left(u^*\right)\right) - f_i\left(F_i(u)\right) > M(u)\left(f_j\left(F_j(u)\right) - f_j\left(F_j\left(u^*\right)\right)\right),$$

for any $M(u) > 0$ and all j satisfying $f_j\left(F_j(u)\right) > f_j\left(F_j\left(u^*\right)\right)$.

Let $A = \left\{ j \in I : f_j\left(F_j(u)\right) > f_j\left(F_j\left(u^*\right)\right) \right\}$, where $I = \{1, \dots, p\}$.

Let $B = I \setminus A \cup \{i\}$. Choose $M(u) > 0$ such that $\dfrac{M(u)}{|A|} > \dfrac{\tau_j}{\tau_i}$, $j \in A$.

Notice that $|L|$ denotes the number of element in the set L. Then

$$\tau_i\left(f_i\left(F_i\left(u^*\right)\right) - f_i\left(F_i(u)\right)\right) > \sum_{j \in A} \tau_j\left(f_j\left(F_j(u)\right) - f_j\left(F_j\left(u^*\right)\right)\right),$$

Since $f_i\left(F_i(u)\right) - f_i\left(F_i\left(u^*\right)\right) > 0$, $\forall\, j \in A$. Therefore,

$$\sum_{i=1}^{p} \tau_i f_i\left(F_i(u)\right) = \tau_i f_i\left(F_i(u)\right) + \sum_{j \in A} \tau_j f_j\left(F_j(u)\right) + \sum_{j \in B} \tau_j f_j\left(F_j(u)\right)$$

$$< \tau_i f_i\left(F_i\left(u^*\right)\right) + \sum_{j \in A} \tau_j f_j\left(F_j\left(u^*\right)\right) + \sum_{j \in B} \tau_j f_j\left(F_j\left(u^*\right)\right)$$

$$= \sum_{i=1}^{p} \tau_i f_i\left(F_i\left(u^*\right)\right).$$

This contradicts the weak duality property. Hence (u, τ, λ) is a conditionally properly efficient solution for (VD).

In the following Theorem it is assumed that f_i, g_j are V − invex and the generalized null space condition (GNC) holds with

$$\delta_i(x, u) = \theta_j(x, u) = 1, \ \forall\, i, j.$$

Theorem 5.4.6: If (u, v) is optimal for (VD$_\tau$) and a dual constraint qualification holds, then u is optimal for (VP$_\tau$).

Proof: Since (u, v) is optimal for the dual problem and a constraint qualification holds at (u, v) then (u, v) satisfies the Kuhn-Tucker conditions:

$$0 \in \sum_{i=1}^{p} \tau_i \partial f_i(F_i(u)) F_i'(u) + \sum_{j=1}^{m} \lambda_j \partial g_j(G_j(u)) G_j'(u) - (C-u)^+$$

$$\lambda_j g_j(G_j(u)) \geq 0, \quad j = 1, \dots, m,$$

$$\lambda_j \geq 0.$$

For any $x \in X$,

$$\sum_{i=1}^{p} \tau_i \left(f_i(F_i(x)) - f_i(F_i(u)) \right)$$

$$\geq \sum_{i=1}^{p} \tau_i \alpha_i(x,u) \xi_i \left(F_i(x) - F_i(u) \right), \forall \xi_i \in \partial f_i(F_i(u))$$

$$= \sum_{i=1}^{p} \tau_i \alpha_i(x,u) \xi_i \eta(x,u) F_i'(u) \mu(x,u)$$

$$\geq -\sum_{j=1}^{m} \lambda_j \beta_j(x,u) \varsigma_j \eta(x,u) G_j'(u) \mu(x,u)$$

$$\geq -\sum_{j=1}^{m} \lambda_j g_j(G_j(x)) + \sum_{j=1}^{m} \lambda_j g_j(G_j(u))$$

$$\geq 0.$$

Therefore, u is an optimal solution for (VD$_\tau$).

The proof of the following Theorem 5.4.7 follows from Theorem 5.4.3 and Theorem 5.4.6.

Theorem 5.4.7: If (u, v) is optimal for (VD$_\tau$) and a constraint qualification holds at (u, v), then (u, v) is conditionally properly efficient solution for (VD) and u is conditionally properly efficient for (VP).

5.5 Lagrange Multipliers and Saddle Point Analysis

The Lagrange multipliers of multiobjective programming problem and the saddle points of its vector-valued Lagrangian function have been studied by many authors e.g., Corley (1987), Craven (1978, 1990), Henig (1982), Jahn (1985), Sawaragi, Nakayama and Tanino (1985), Tanaka (1988, 1990), Vogel (1974), Wang (1984), and Weir,Mond and Craven (1986, 1987). However, in most of the studies an assumption of convexity on the problems was made.

In this Section, we extend the relevant results using $V-$ invex functions and its generalizations. As a consequence of Theorem 5.3.1, a Lagrange multipliers theorem is established and vector valued saddle point results are also obtained. The results of this Section and that of the next Section have appeared in Mishra (1996).

Theorem 5.5.1: If Theorem 5.3.1 holds, then equivalent multiobjective composite problem (EVP) for (VP) 9s given by (EVP)

$$V-\underset{x\in C}{\text{Minimize}}\left(f_1(F_1(x))+\lambda^T g(G(x)),...,f_p(F_p(x))+\lambda^T g(G(x))\right)$$

$$\text{subject to}\quad \lambda_j g_j(G_j(x))=0,\quad j=1,...,m$$

$$\lambda_j\geq 0,\quad j=1,...,m.$$

Proof: Let x^0 be a Pareto optimum for (VP), from the optimality conditions (KT), we have

$$0\in\sum_{i=1}^{p}\tau_i\partial f_i(F_i(u))F_i'(u)+\sum_{j=1}^{m}\lambda_j\partial g_j(G_j(u))G_j'(u)-(C-u)^+$$

$$\lambda_j g_j(G_j(u))=0,\quad j=1,...,m.$$

Therefore, we have

$$0\in\sum_{i=1}^{p}\tau_i\left\{\partial f_i(F_i(u))F_i'(u)+\lambda^T g(G(u))\right\}$$

$$+\sum_{j=1}^{m}\lambda_j\partial g_j(G_j(u))G_j'(u)-(C-u)^+$$

Now applying the arguments of Theorem 5.3.1 by replacing $f_i(F_i(x))$ by $f_i(F_i(x))+\lambda^T g(G(x))$ yields the result.

Theorem 5.5.1: suggests the vector valued Lagrangian function $L(x,\lambda)$ as $L:C\times R_+^m\to R^p$ given by

$$L(x,\lambda)=(L_1(x,\lambda),...,L_p(x,\lambda)),$$

where $L_i(x,\lambda)=f_i(F_i(x))+\lambda^T g(G(x)),\quad i=1,...,p.$

Definition 5.5.1: A point $(x^0,\lambda^0)\in C\times R_+^m$ is said to be a vector saddle point of the vector valued Lagrangian function $L(x,\lambda)$ if it satisfies the following conditions

$$L\left(x^0, \lambda\right) \not\geq L\left(x^0, \lambda^0\right), \qquad \forall \lambda \in R_+^m \tag{5.1}$$

and

$$L\left(x^0, \lambda^0\right) \not\geq L\left(x, \lambda^0\right), \qquad \forall x \in C. \tag{5.2}$$

Theorem 5.5.2: If $\left(x^0, \lambda^0\right)$ is a vector saddle point of $L(x, \lambda)$, then x^0 is a conditionally properly efficient solution for (VP).

Proof: Since $\left(x^0, \lambda^0\right)$ is a vector saddle point of $L(x, \lambda)$, therefore, we have $L_i\left(x^0, \lambda\right) \leq L_i\left(x^0, \lambda^0\right)$,   for atleast one i and $\forall \lambda \in R_+^m$

$$\Rightarrow f_i\left(F_i\left(x^0\right)\right) + \lambda^T g\left(G\left(x^0\right)\right) \leq f_i\left(F_i\left(x^0\right)\right) + \lambda^{0^T} g\left(G\left(x^0\right)\right),$$

for atleast one i and $\forall \lambda \in R_+^m$

$$\Rightarrow \left(\lambda - \lambda^0\right)^T g\left(G\left(x^0\right)\right) \leq 0, \qquad \forall \lambda \in R_+^m.$$

This gives $g\left(G\left(x^0\right)\right) \leq 0$.

First we show that x^0 is an efficient solution for (VP). Since x^0 is feasible for (VP), we have $\lambda^{0^T} g\left(G\left(x^0\right)\right) \leq 0$. But, by setting $\lambda = 0$, then from $\left(\lambda - \lambda^0\right)^T g\left(G\left(x^0\right)\right) \leq 0$, we get $\lambda^{0^T} g\left(G\left(x^0\right)\right) \geq 0$. Thus

$$\lambda^{0^T} g\left(G\left(x^0\right)\right) = 0.$$

Assume contrary, i.e., x^0 is not an efficient solution for (VP). Therefore, there exists an $x \in C$ with $g\left(G(x)\right) \leq 0$ such that

$$f_i\left(F_i(x)\right) \leq f_i\left(F_i\left(x^0\right)\right), \quad \forall i = 1, \dots, p \text{ and } \forall x \in C$$

and

$$f_k\left(F_k(x)\right) < f_k\left(F_k\left(x^0\right)\right), \qquad \text{for atleast one } k \text{ and } \forall \lambda^0 \in R_+^m.$$

These along with $\lambda^{0^T} g\left(G\left(x^0\right)\right) = 0$ yields

$$f_i\left(F_i(x)\right) + \lambda^{0^T} g\left(G\left(x^0\right)\right) \leq f_i\left(F_i\left(x^0\right)\right) + \lambda^{0^T} g\left(G\left(x^0\right)\right) = 0,$$

$$\forall i = 1, \dots, p \text{ and } \forall x \in C$$

and

$$f_k\left(F_k(x)\right) + \lambda^{0^T} g\left(G\left(x^0\right)\right) < f_k\left(F_k\left(x^0\right)\right) + \lambda^{0^T} g\left(G\left(x^0\right)\right),$$

for at least one k and $\forall \lambda^0 \in R_+^m$.

That is,

$$L_i\left(x,\ \lambda^0\right)\le L_i\left(x^0,\ \lambda^0\right), \qquad \forall\ \ i=1,...,p \ \text{ and } \ \forall\ x\in C\ ,$$

and

$$L_k\ x,\ \lambda^0\ <L_k\ x^0,\ \lambda^0\ , \qquad \text{for atleast one } k \text{ and } \forall\ \lambda^0\in R_+^m\ ,$$

which is a contradiction to (5.2). Hence x^0 is an efficient solution for (VP).

We now suppose that x^0 is not a conditionally properly efficient solution for (VP). Therefore, there exists a feasible point x for (VP) and an index i such that for every positive function $M\left(x^0\right)>0$, we have

$$\frac{f_i\left(F_i\left(x\right)\right)-f_i\left(F_i\left(x^0\right)\right)}{f_j\left(F_j\left(x^0\right)\right)-f_j\left(F_j\left(x\right)\right)}>M\left(x^0\right),$$

for all j satisfying $f_j\left(F_j\left(x^0\right)\right)<f_j\left(F_j\left(x\right)\right),$ whenever

$$f_i\left(F_i\left(x^0\right)\right)>f_i\left(F_i\left(x\right)\right).$$

This along with $\lambda^{0^T}g\left(G\left(x^0\right)\right)=0$ and $\lambda^{0^T}g\left(G\left(x\right)\right)\le 0$ yields

$$f_i\left(F_i(x)\right)+\lambda^{0^T}g\left(G(x)\right)<f_i\left(F_i(x^0)\right)+\lambda^{0^T}g\left(G(x^0)\right),$$

$$\forall i=1,...,p \text{ and } \forall x\in C$$

which is a contradiction to (5.2). Hence x^0 is a conditionally properly efficient solution for (VP).

Theorem 5.5.3: Let x^0 be a conditionally properly efficient solution for (VP) and let at x^0 Slater type constraint qualification be satisfied. If $\left(f_1,...,f_p\right)$ and $\left(-g_1,...,-g_p\right)$ are $V-$ invex on the set C and $F_i,\ i=1,...,p$ and $G_j,\ j=1,...,m$ are locally Lipschitz and Gateaux differentiable functions. Then there exists $\lambda^0\in R_+^m$ such that $\left(x^0,\ \lambda^0\right)$ is a vector saddle point of $L\left(x,\ \lambda\right)$.

Proof: Since x^0 is a conditionally properly efficient solution for (VP), therefore, x^0 is also an efficient solution for (VP) and since at x^0 Slater type constraint qualification is satisfied, therefore, by Theorem 5.3.1, there exist $\tau^0\in R^p$ with $\tau^0>0$ and $\lambda^0\in R_+^m$ such that the following hold:

$$0\in\sum_{i=1}^{p}\tau_i\partial f_i\left(F_i\left(x^0\right)\right)F_i'\left(x^0\right)+\sum_{j=1}^{m}\lambda_j^0\partial g_j\left(G_j\left(x^0\right)\right)G_j'\left(x^0\right)-\left(C-x^0\right)^+\ , \quad (5.3)$$

$$\lambda_j^0 g_j G_j\left(x^0\right)=0, \quad j=1,...,m. \tag{5.4}$$

These yields

$$\sum_{i=1}^{p} \tau_i \xi_i F_i'(x^0) + \sum_{j=1}^{m} \lambda_j^0 \varsigma_j G_j'(x^0) + z^0 = 0, \qquad (5.5)$$

for some $\xi_i \in \partial f_i\left(F_i(x^0)\right)$, $\quad i = 1,...,p$ and $\varsigma_j \in \partial g_j\left(G_j(x^0)\right)$, $j = 1,...,m$ and $z^0 \in \left(C - x^0\right)^+$.

Using the V − invexity assumption of the functions, we obtain

$$f_i\left(F_i(x)\right) - f_i\left(F_i(x^0)\right) \geq \alpha_i(x, x^0)\xi_i \eta(x, x^0)F_i'(x^0)\mu(x, x^0),$$

$$\forall i = 1,...,p \text{ and } \forall x \in C$$

and

$$f_k\left(F_k(x)\right) - f_k\left(F_k(x^0)\right) \geq \alpha_k(x, x^0)\xi_k \eta(x, x^0)F_k'(x^0)\mu(x, x^0),$$

for at least one k and $\forall x \in C$. Since $\alpha_i\left(x, x^0\right) > 0$, $\quad \forall \; i = 1,...,p$, we get

$$f_i\left(F_i(x)\right) - f_i\left(F_i(x^0)\right) \geq \xi_i \eta(x, x^0)F_i'(x^0)\mu(x, x^0), \qquad (5.6)$$

$$\forall i = 1,...,p \text{ and } \forall x \in C$$

and

$$f_k\left(F_k(x)\right) - f_k\left(F_k(x^0)\right) \geq \xi_k \eta(x, x^0)F_k'(x^0)\mu(x, x^0), \qquad (5.7)$$

for at least one k and $\forall x \in C$.

Now for all $i = 1,...,p$, $\forall \; x \in C$, we have

$$L_i(x, \lambda^0) - L_i(x^0, \lambda^0)$$

$$= f_i\left(F_i(x)\right) - f_i\left(F_i(x^0)\right) + \lambda^{0^T}\left[g\left(G(x)\right) - g\left(G(x^0)\right)\right]$$

$$\geq -\eta(x, x^0)\mu(x, x^0)\left[\sum_{j=1}^{m} \lambda_j^0 \varsigma_j G_j'(x^0) - \left(C - x^0\right)^+\right]$$

$$\geq 0 \text{ (because } g_j, \; j = 1,...,m \text{ are } V - \text{invex and } z^0 \in \left(C - x^0\right)^+).$$

Since $\tau \in R^p$, $\tau_i > 0$, $i = 1,...,p$, therefore,

$$L_i\left(x^0, \lambda^0\right) \not\geq L_i\left(x, \lambda^0\right), \quad \forall \; x \in C.$$

The other part
$L_i\left(x^0, \lambda\right) \not\geq L_i\left(x^0, \lambda^0\right)$, $\forall \; \lambda \in R_+^m$ of the vector saddle point inequality follows from

$$L\left(x^0, \lambda\right) - L\left(x^0, \lambda^0\right) = \left(\lambda - \lambda^0\right)^T g\left(G(x^0)\right) \leq 0.$$

Hence $\left(x^0, \lambda^0\right)$ is a vector saddle point of $L(x, \lambda)$.

Remark 5.5.1: Theorem 5.5.3 can be established under weaker V – invexity assumptions, namely, $\left(\tau_1 f_1, \ldots, \tau_p f_p\right)$ and $\left(-\tau_1 g_1, \ldots, -\tau_p g_p\right)$ are V – pseudo-invex and $\left(\lambda_1 h_1, \ldots, \lambda_m h_m\right)$ is V – quasi-invex.

5.6 Scalarizations in Composite Multiobjective Programming

In this Section, we present a scalarization result for nonconvex composite problems. As an application of the scalarization result we also characterize the set of conditionally properly efficient solutions in terms of subgradients (Rockafellar (1969)) for V – invex problems. These conditions do not depend on a particular conditionally properly efficient solution, and differ from the conditions presented in Mishra and Mukherjee (1995a).

For the multiobjective composite problem

(VP) V – Minimize $\left(f_1\left(F_1(x)\right), \ldots, f_p\left(F_p(x)\right)\right)$

subject to $x \in C$, $g_j\left(G_j(x)\right) \le 0$, $j = 1, \ldots, m$,

The associated scalar problem

(VP$_\tau$)   Minimize $\sum_{i=1}^{p} \tau_i f_i\left(F_i(x)\right)$

subject to $x \in C$, $\lambda_j g_j\left(G_j(x)\right) \le 0$, $j = 1, \ldots, m$,

where $\tau \in R^p$, $\tau \ne 0$. The feasible set Ω for (VP) is given by

$$\Omega = \left\{x \in C : g_j\left(G_j(x)\right) \le 0, \quad j = 1, \ldots, m\right\}.$$

The set of all conditionally properly efficient solutions for (VP) is denoted by CPE. For each $\tau \in R^p$, the solution set S_τ of the scalar problem (VP$_\tau$) is given by

$$S_\tau = \left\{x \in \Omega : \sum_{i=1}^{p} \tau_i f_i\left(F_i(x)\right) = \min_{y \in \Omega} \sum_{i=1}^{p} \tau_i f_i\left(F_i(y)\right)\right\}.$$

The following Theorem establish a scalarization result for (VP) corresponding to a conditionally properly efficient solution.

Theorem 5.6.1: For the multiobjective problem (VP), assume that, for each $i = 1, ..., p$ and $\alpha > 0$ the set

$$\Gamma_\alpha^i = \{z \in R^p : \exists x \in C, f_i(F_i(x)) < z_i^i,$$
$$f_i(F_i(x)) + \alpha f_j(F_j(x)) < z_i^j, j \neq i\}$$

is convex. Then, CPE $= \displaystyle\bigcup_{\tau_i > 0, \sum_{i=1}^{p} \tau_i = 1} S_\tau$

Proof: Let $u \in \text{CPE}$. Then, there exists a positive function $M(u) > 0$ such that, for each $i = 1, ..., p$, the system $f_i(F_i(x)) < f_i(F_i(u))$,

$$f_i(F_i(x)) + M(u)f_j(F_j(x)) < f_i(F_i(u)) + M(u)f_j(F_j(u)), \quad \forall \ j \neq i,$$

is inconsistent.

Thus,

$$0 \notin \Gamma_{M(u)}^i(u) = \left\{ \begin{array}{l} z \in R^p : \exists x \in C, f_i(F_i(x)) < f_i(F_i(u)) + z_i^i, \\ f_i(F_i(x)) + M(u)f_j(F_j(x)) < \\ f_i(F_i(x)) + M(u)f_j(F_j(u)) + z_i^j, j \neq i \end{array} \right\}.$$

From the assumption, $\Gamma_{M(u)}^i(u)$ is convex, now on the lines of the proof of Theorem 5.1 of Jeyakumar and Yang (1993), we can show that there exists $\tau \in R^p$, $\tau_i > 0$, $\displaystyle\sum_{i=1}^{p} \tau_i = 1$ such that $u \in S_\tau$; thus,

$$\text{CPE} = \bigcup_{\tau_i > 0, \ \sum_{i=1}^{p} \tau_i = 1} S_\tau.$$

The converse inclusion follows as in the proof of Theorem 5.1 of Jeyakumar and Yang (1993) without any convexity conditions on the functions involved.

Using the above scalarization Theorem 5.6.1 and a result of Mangasarian (1988) we show how the set of conditionally properly efficient solutions for a nonconvex problem can be characterized in terms of subgradients. This extends the characterization result of Mangasarian (see Theorem 1(a), Mangasarian (1988)) and that of Jeyakumar and Yang (see Corollary 5.1, Jeyakumar and Yang (1993)) for a scalar problem to multiobjective nonconvex problems. In the following, we assume that the functions F_i, $i = 1, ..., p$ and G_j, $j = 1, ..., m$ in problem (VP) are linear functions from R^n and R^m, respectively. Thus, we consider the composite nonconvex problem

(NCP) $V - \text{Minimize } \left(f_1(A_1(x)),..., f_p(A_p(x)) \right)$

subject to $x \in C, \ g_j\left(B_j(x)\right) \leq 0, \ j = 1,...,m,$

where $A_i : R^n \to R^m, \ i = 1,...,p$ and $B_j : R^n \to R^m, \ j = 1,...,m$ are continuous linear mappings, $f_i : R^m \to R, \ i = 1,...,p$ and $g_j : R^m \to R, \ j = 1,...,m$ are convex functions on R^m. Note that the feasible set

$$\Omega = \left\{ x \in C : g_j\left(B_j(x)\right) \leq 0, \quad j = 1,...,m \right\}$$

is now a convex subset of R^n.

The nonconvex scalar problem for (NCP) is given by

(NCP$_\tau$) $\text{Minimize } \displaystyle\sum_{i=1}^{p} \tau_i \ f_i\left(A_i(x)\right)$

subject to $x \in C, \ g_j\left(B_j(x)\right) \leq 0, \ j = 1,...,m,$

Let the convex solution set of (NCP$_\tau$) be $CS_\tau, \ \tau \in R^p$.

Corollary 5.6.1: Consider the nonconvex problem (NCP). Suppose that for each $\tau \in R^p, \tau_i > 0, \displaystyle\sum_{i=1}^{p} \tau_i = 1$, the relative interior of CS_τ, $\left(\text{ri}\left(CS_\tau\right)\right)$, is non-empty. Let $z_\tau \in \text{ri}\left(CS_\tau\right)$. Then

$$\text{CPE} = \bigcup_{\tau_i > 0, \ \sum_{i=1}^{p} \tau_i = 1} \left\{ x \in \Omega : \exists \ u_i \in \partial f_i\left(A_i(x)\right), \ \sum_{i=1}^{p} \tau_i u_i^T A_i\left(x - z_\lambda\right) = 0 \right\}.$$

Proof: Proof of this Corollary follows the lines of the proof of Corollary 5.1 of Jeyakumar and Yang (1993).

Chapter 6: Continuous-time Programming

6.1 Introduction

The optimization problems in the previous Chapters have all been finite dimensional and functions have been defined on R^n and the number of constraints has been finite. However, a great deal of optimization theory is concerned with problems involving infinite dimensional normal spaces.

Two types of problems fitting into this scheme are variational and control problems. An early result of Friedrichs (1929) for a simple variational problem has been presented by Courant and Hilbert (1948). Hanson (1964) observed that variational and control problems are continuous analogues of finite dimensional nonlinear programs. Since, then the fields of nonlinear programming and the calculus of variations have to some extent, merged together within optimization theory, enhancing the potential for continued research in both. In particular, Mond and Hanson (1967, 1968) gave duality theorems for variational and control problems using convexity assumptions. Chandra, Craven and Husain (1985) established optimality conditions and duality results for a class of continuous programming problems with a nondifferentiable term in the integrand of the objective function. Mond, Chandra and Husain (1988) extended the concept of invexity to continuous functions. Mond and Smart (1988) established duality results using invexity assumptions and proved that the necessary conditions for optimality in the control problems are also sufficient . Mishra and Mukherjee (1994b) obtained various duality results for multiobjective variational problems. See also Kim and Kim (2002), Kim and Lee (1998), Kim et al. (1998), Kim et al. (2004).

Mond and Husain (1989) obtained a number of Kuhn-Tucker type sufficient optimality criteria for a class of variational problems under weaker invexity assumptions. As an application of these optimality results, various Mond-Weir type duality results are proved under a variety of generalized invexity assumptions. These results generalize many well known duality results of variational problems and also give a dynamic analogue to certain corresponding (static) results relating to duality with generalized invexity in mathematical programming.

In this Chapter, we extend the concept of V – invexity to continuous functions and to continuous functionals and use it to obtain sufficient optimality conditions and duality results for different kinds of multiobjective variational and control problems. For this purpose the Chapter is divided in six sections. In Section 2, we extend the concept of V – invexity to continuous functions and discuss some examples. In Section 3, we present a number of Kuhn-Tucker type sufficient optimality conditions. In Section 4, Mond-Weir type duality results are obtained under a variety of V – invexity assumptions. In Section 5, we have presented multiobjective control problems and obtained duality theorems. In last Section, we have considered a class of nondifferentiable multiobjective variational problems and establish duality results mainly for conditionally properly efficient solutions of the problem.

6.2 V – Invexity for Continuous-time Problems

Let $I = [a, b]$ be a real interval and $\psi : I \times R^n \times R^n \to R$ be a continuously differentiable function. In order to consider $\psi(t, x, \dot{x})$, where $x : I \to R^n$ is differentiable with derivative $\dot{x}$, we denote the partial derivatives of ψ by ψ_t, $\psi_x = \left[\dfrac{\partial \psi}{\partial x^1}, ..., \dfrac{\partial \psi}{\partial x^n} \right]$, $\psi_{\dot{x}} = \left[\dfrac{\partial \psi}{\partial \dot{x}^1}, ..., \dfrac{\partial \psi}{\partial \dot{x}^n} \right]$. The partial derivatives of the other functions used will be written similarly. Let X denote the space of piecewise smooth functions x with norm $\|x\| = \|x\|_\infty + \|Dx\|_\infty$, where the differential operator D is given by

$$u^i = Dx^i \Leftrightarrow x^i(t) = \alpha + \int_a^t u^i(s)\, ds,$$

where α is a given boundary value. Therefore, $D = \dfrac{d}{dt}$ except at discontinuities.

Let $F_i : X \to R$ defined by $F_i(x) = \int_a^b f_i(t, x, \dot{x})\, dt$, $i = 1, ..., p$ be differentiable. The following definitions and examples have appeared in Mukherjee and Mishra (1994).

Definition 6.2.1 ($V-$**Invex**): A vector function $F = (F_1,...,F_p)$ is said to be $V-$ invex if there exists differentiable vector function $\eta : I \times R^n \times R^n \to R^n$ with $\eta(t,x,x)=0$ and $\beta_i : I \times X \times X \to R_+ \setminus \{0\}$ such that for each $x, \overline{x} \in X$ and for $i = 1,...,p$

$$F_i(x) - F_i(\overline{x}) \geq \int_a^b \left[\alpha_i(t,x(t),\overline{x}(t)) f_x^i(t,x(t),\dot{x}(t)) \eta(t,x(t),\overline{x}(t)) \right.$$

$$\left. + \frac{d}{dt} \eta(t,x(t),\overline{x}(t)) \alpha_i(t,x(t),\overline{x}(t)) f_{\dot{x}}^i(t,x(t),\dot{x}(t)) \right] dt.$$

Definition 6.2.2 ($V-$**Pseudo-Invex**):

A vector function $F = (F_1,...,F_p)$ is said to be $V-$ pseudo-invex if there exists differentiable vector function $\eta : I \times R^n \times R^n \to R^n$ with $\eta(t,x,x)=0$ and $\beta_i : I \times X \times X \to R_+ \setminus \{0\}$ such that for each $x, \overline{x} \in X$ and for $i = 1,...,p$

$$\int_a^b \left[\sum_{i=1}^p \eta(t,x,\overline{x}) f_x^i(t,x,\dot{x}) + \frac{d}{dt} \eta(t,x,\overline{x}) f_{\dot{x}}^i(t,x,\dot{x}) \right] dt \geq 0$$

$$\Rightarrow \int_a^b \left[\sum_{i=1}^p \beta_i(t,x(t),\overline{x}(t)) f_i(t,x(t),\dot{x}(t)) \right] dt$$

$$\geq \int_a^b \left[\sum_{i=1}^p \beta_i(t,x(t),\overline{x}(t)) f_i(t,x(t),\dot{x}(t)) \right] dt$$

Or equivalently;

$$\int_a^b \left[\sum_{i=1}^p \beta_i(t,x(t),\overline{x}(t)) f_i(t,x(t),\dot{x}(t)) \right] dt$$

$$< \int_a^b \left[\sum_{i=1}^p \beta_i(t,x(t),\overline{x}(t)) f_i(t,x(t),\dot{x}(t)) \right] dt$$

$$\Rightarrow \int_a^b \left[\sum_{i=1}^p \eta(t,x,\overline{x}) f_x^i(t,x,\dot{x}) + \frac{d}{dt} \eta(t,x,\overline{x}) f_{\dot{x}}^i(t,x,\dot{x}) \right] dt < 0.$$

Definition 6.2.3 ($V-$**Quasi-Invex**):

A vector function $F = \left(F_1, ..., F_p \right)$ is said to be V – quasi-invex if there exists differentiable vector function $\eta : I \times R^n \times R^n \rightarrow R^n$ with $\eta(t, x, x) = 0$ and $\beta_i : I \times X \times X \rightarrow R_+ \setminus \{0\}$ such that for each $x, \overline{x} \in X$ and for $i = 1, ..., p$

$$\int_a^b \left[\sum_{i=1}^p \beta_i \left(t, x(t), \overline{x}(t) \right) f_i \left(t, x(t), \dot{\overline{x}}(t) \right) \right] dt$$

$$\leq \int_a^b \left[\sum_{i=1}^p \beta_i \left(t, x(t), \overline{x}(t) \right) f_i \left(t, x(t), \dot{\overline{x}}(t) \right) \right] dt$$

$$\Rightarrow \int_a^b \left[\sum_{i=1}^p \eta(t, x, \overline{x}) f_x^i \left(t, x, \dot{\overline{x}} \right) + \frac{d}{dt} \eta(t, x, \overline{x}) f_{\dot{x}}^i \left(t, x, \dot{\overline{x}} \right) \right] dt \leq 0;$$

Or equivalently;

$$\int_a^b \left[\sum_{i=1}^p \eta(t, x, \overline{x}) f_x^i \left(t, x, \dot{\overline{x}} \right) + \frac{d}{dt} \eta(t, x, \overline{x}) f_{\dot{x}}^i \left(t, x, \dot{\overline{x}} \right) \right] dt > 0$$

$$\Rightarrow \int_a^b \left[\sum_{i=1}^p \beta_i \left(t, x(t), \overline{x}(t) \right) f_i \left(t, x(t), \dot{\overline{x}}(t) \right) \right] dt$$

$$> \int_a^b \left[\sum_{i=1}^p \beta_i \left(t, x(t), \overline{x}(t) \right) f_i \left(t, x(t), \dot{\overline{x}}(t) \right) \right] dt$$

It is to be noted here that, if the function f is independent of t, Definitions 6.2.1-6.2.3 reduce to the definitions of V – invexity, V – pseudo-invexity and V – quasi-invexity of Jeyakumar and Mond (1992), respectively and given in Chapter 2. It is apparent that every V – invex function is V – pseudo-invex and V – quasi-invex.

The following example shows that V – invexity is wider than that of invexity:

Example 6.2.1: Consider

$$\min_{x_1, x_2 \in R} \left(\int_a^b \frac{x_1^2(t)}{x_2(t)} dt, \int_a^b \frac{x_2(t)}{x_1(t)} dt \right)$$

subject to $1 - x_1(t) \leq 0, \quad 1 - x_2(t) \leq 0.$

Then for

$$\alpha_1(x, u) = \frac{u_2(t)}{x_2(t)}, \quad \alpha_2(x, u) = \frac{u_1(t)}{x_1(t)}, \quad \beta_i(x, u) = 1, \quad \text{for} \quad i = 1, 2 \text{ and}$$

$\eta(x, u) = x(t) - u(t)$. We shall show that

$$\int_a^b f_i(t, x, \dot{x}) - f(t, u, \dot{u}) -$$

$$\alpha_i(t, x(t), u(t)) f_x^i(t, x(t), u(t)) \eta(t, x(t), u(t)) dt \geq 0, \quad i = 1, 2.$$

Now,

$$\int_a^b \frac{x_1^2(t)}{x_2(t)} dt - \int_a^b \frac{u_1^2(t)}{u_2(t)} dt - \int_a^b \frac{u_2(t)}{x_2(t)} \left(\frac{2u_1(t)}{u_2(t)}, \frac{-u_1^2(t)}{u_2^2(t)} \right) (x_1 - 1)(x_2 - 1) dt$$

$$= \int_a^b \frac{x_1^2(t)}{x_2(t)} dt - \int_a^b 1 \, dt - \int_a^b \frac{1}{x_2(t)} (2, -1)(x_1 - 1)(x_2 - 1) dt$$

$$= \int_a^b \frac{x_1^2(t)}{x_2(t)} dt - \int_a^b 1 \, dt - \int_a^b \frac{1}{x_2(t)} (2x_1 - 2 - \bar{x}_2 + 1) dt$$

$$= \int_a^b \frac{x_1^2(t)}{x_2(t)} dt - \int_a^b 1 \, dt - \int_a^b \left\{ \frac{2x_1}{x_2(t)} - 1 - \frac{1}{x_2(t)} \right\} dt$$

$$= \int_a^b \left\{ \frac{x_1^2(t)}{x_2(t)} - \frac{2x_1}{x_2(t)} + \frac{1}{x_2(t)} \right\} dt$$

$$= \int_a^b \left\{ \frac{(x_1(t) - 1)^2}{x_2(t)} \right\} dt \geq 0.$$

The following example shows that V − invex functions can be formed from certain nonconvex functions.

Example 6.2.2: Consider the function $h : I \times X \times X \to R^p$

$$h(t, x(t), \dot{x}(t)) = \left(\int_a^b f_1(t, x(t), \dot{x}(t)) dt, \dots, \int_a^b f_p(t, x(t), \dot{x}(t)) dt \right)$$

where $f_i : I \times X \times X \to R$, $i = 1, \dots, p$ are strongly pseudo-convex functions with real positive functions $\alpha_i(t, x, u)$, $\psi : I \times X \times X \to R^n$ is surjective with $\psi'(t, u, \dot{u})$ onto for each $u \in R^n$. Then the function

$h : I \times X \times X \to R^p$ is $V-$ invex. To see this, let $x, u \in X$, $v = \psi(t, x, \dot{x})$, $w = \psi(t, u, \dot{u})$. Then, by strong-pseudo-convexity, we get

$$\int_a^b \{f_i(\psi(t, x, \dot{x})) - f_i(\psi(t, u, \dot{u}))\} dt = \int_a^b \{f_i(v) - f_i(w)\} dt$$

$$\geq \int_a^b \alpha_i(t, v, w) f_i'(w)(v - w)\psi_x(t, x, \dot{x}) dt$$

$$+ \int_a^b \frac{d}{dt}\alpha_i(t, v, w)(v - w) f_i'(w)\psi_{\dot{x}}(t, x, \dot{x}) dt.$$

Since $\psi'(t, u, \dot{u})$ is onto, $v - w = \psi'(t, u, \dot{u})\eta(t, x, u)$ is solvable for some $\eta(t, x, u)$.

Hence

$$\int_a^b \{f_i(\psi(t, x, \dot{x})) - f_i(\psi(t, u, \dot{u}))\} dt \geq \int_a^b \alpha_i(t, v, w)(f_i \circ \psi)_x dt$$

$$+ \int_a^b \frac{d}{dt}\alpha_i(t, v, w)\eta(t, v, w)(f_i \circ \psi)_x dt$$

Now consider the determination of a piecewise smooth extremal $x = x(t)$, $a \leq t \leq b$, for the following multiobjective variational problem:

(VCP)   Minnimize $\int_a^b f(t, x, \dot{x}) dt = \left(\int_a^b f_1(t, x, \dot{x}) dt, ..., \int_a^b f_p(t, x, \dot{x}) dt \right)$

subject to

$$x(a) = \alpha, \quad x(b) = \beta \tag{6.1}$$

$$g(t, x, \dot{x}) \leq 0, \quad t \in I. \tag{6.2}$$

where $f_i : I \times R^n \times R^n \to R$, $i \in P = \{1, ..., p\}$, $g : I \times R^n \times R^n \to R^m$ are assumed to be continuously differentiable functions.

Let K be the set of all feasible solutions for (VCP), that is,

$$K = \{x \in X : x(a) = \alpha, x(b) = \beta, g(t, x(t), \dot{x}(t)) \leq 0, t \in I\}.$$

Consider also the determination of $m + n$ dimensional extremal $(u, \lambda) = (u(t), \lambda(t))$, $t \in I$, for the following maximization problem:

(VCD) $V - Maximize \int_a^b f(t,u,\dot{u})dt = \left(\int_a^b f_1(t,u,\dot{u})dt,...,\int_a^b f_p(t,u,\dot{u})dt\right)$

subject to
$$u(a) = \alpha, \quad u(b) = \beta, \tag{6.3}$$

$$\sum_{i=1}^p \tau_i f_u^i(t,u,\dot{u}) + \sum_{j=1}^m \lambda_j g_u^j(t,u,\dot{u}) \tag{6.4}$$

$$= \frac{d}{dt}\left(\sum_{i=1}^p \tau_i f_{\dot{u}}^i(t,u,\dot{u}) + \sum_{j=1}^m \lambda_j(t) g_{\dot{u}}^j(t,u,\dot{u})\right),$$

$$\int_a^b \lambda_j(t) g_j(t,u,\dot{u}) dt \geq 0, \quad t \in I, \quad j = 1,...,m, \tag{6.5}$$

$$\lambda(t) \geq 0, \ t \in I, \qquad \tau e = 1, \quad \tau \geq 0, \tag{6.6}$$

where $e = (1,...,1) \in R^p$.

6.3 Necessary and Sufficient Optimality Criteria

In this section, we present sufficient optimality criteria of the Kuhn-Tucker type for the problem (VCP). The following necessary optimality conditions will be shown to be sufficient for optimality under generalized V − invexity assumptions. There exists a piecewise smooth $\lambda^* : I \to R^m$ such that

$$\sum_{i=1}^p \tau_i f_x^i(t,x^*,\dot{x}^*) + \sum_{j=1}^m \lambda_j^*(t) g_x^j(t,x^*,\dot{x}^*) \tag{6.7}$$

$$= \frac{d}{dt}\left(\sum_{i=1}^p \tau_i f_{\dot{x}}^i(t,x^*,\dot{x}^*) + \sum_{j=1}^m \lambda_j^*(t) g_{\dot{x}}^j(t,x^*,\dot{x}^*)\right),$$

$$\lambda_j^*(t) g_x^j(t,x^*,\dot{x}^*) = 0, \ t \in I, \quad j = 1,...,m, \tag{6.8}$$

$$\tau \in R^p, \ \tau \neq 0, \ \tau \geq 0, \quad \lambda^*(t) \geq 0, \ t \in I. \tag{6.9}$$

Theorem 6.3.1 (Sufficient Optimality Conditions):
Let x^* be a feasible solution for (VCP) and assume that

$$\left(\int_a^b \tau_1 f_1(t,\cdot,\cdot) dt, \ldots, \int_a^b \tau_p f_p(t,\cdot,\cdot) dt \right)$$

is V − pseudo-invex and

$$\left(\int_a^b \lambda_1 g_1(t,\cdot,\cdot) dt, \ldots, \int_a^b \tau_m g_m(t,\cdot,\cdot) dt \right)$$

is V − quasi-invex with respect to η. If there exists a piecewise smooth $\lambda^* : I \to R^m$ such that $(x^*(t), \lambda^*(t))$ satisfies the conditions (6.7)-(6.9), then x^* is a global weak minimum for (VCP).

Proof: Suppose that x^* is not a global weak minimum point. Then there exists feasible $x_0 \in X$ such that

$$\int_a^b f_i(t,x_0(t),\dot{x}_0(t)) dt < \int_a^b f_i(t,u(t),\dot{u}(t)) dt, \quad t \in I, \quad i=1,\ldots,p.$$

Therefore,

$$\int_a^b \sum_{i=1}^p \beta_i(t,x_0(t),u(t)) \tau_i f_i(t,x_0(t),\dot{x}_0(t)) dt$$

$$< \int_a^b \sum_{i=1}^p \beta_i(t,x_0(t),u(t)) \tau_i f_i(t,u(t),\dot{u}(t)) dt$$

Now, by the V − pseudo-invexity condition, we get

$$\int_a^b \left[\sum_{i=1}^p \tau_i \eta(t,x_0(t),u(t)) f_x^i(t,u(t),\dot{u}(t)) \right. \tag{6.10}$$

$$\left. + \frac{d}{dt}\eta(t,x_0(t),u(t)) f_{\dot{x}}^i(t,u(t),\dot{u}(t)) \right] dt < 0$$

From (6.7), we have

$$\int_a^b \eta(t,x_0(t),u(t)) \left[\sum_{i=1}^p \tau_i f_x^i(t,u,\dot{u}) + \sum_{j=1}^m \lambda_j^*(t) g_x^j(t,u,\dot{u}) \right] dt$$

$$= \int_a^b \eta(t,x_0(t),u(t)) \frac{d}{dt}\left(\sum_{i=1}^p \tau_i f_{\dot{x}}^i(t,u,\dot{u}) + \sum_{j=1}^m \lambda_j^*(t) g_{\dot{x}}^j(t,u,\dot{u}) \right) dt$$

$$= \eta(t,x_0(t),u(t)) \left(\sum_{i=1}^p \tau_i f_{\dot{x}}^i(t,u,\dot{u}) + \sum_{j=1}^m \lambda_j^*(t) g_{\dot{x}}^j(t,u,\dot{u}) \right) \Bigg|_a^b$$

$$-\int_a^b \frac{d}{dt}\eta(t,x_0(t),u(t))\left(\sum_{i=1}^p \tau_i f_x^i(t,u,\dot{u})+\sum_{j=1}^m \lambda_j^*(t)g_x^j(t,u,\dot{u})\right)dt$$

(integration by part).

Thus,

$$\int_a^b \eta(t,\,x_0(t),u(t))\left[\sum_{i=1}^p \tau_i f_x^i(t,u,\dot{u})+\sum_{j=1}^m \lambda_j^*(t)g_x^j(t,u,\dot{u})\right]dt \qquad (6.11)$$

$$+\int_a^b \frac{d}{dt}\eta(t,x_0(t),u(t))\left(\sum_{i=1}^p \tau_i f_x^i(t,u,\dot{u})+\sum_{j=1}^m \lambda_j^*(t)g_{x'}^j(t,u,\dot{u})\right)dt = 0$$

(Since $\eta(t,u,u)=0$).

From (6.11), we have

$$\int_a^b \sum_{j=1}^m \Big[\eta\big(t,x_0(t),u(t)\big)\lambda_j^*(t)g_x^j\big(t,u,\dot{u}\big) \qquad (6.12)$$

$$+\frac{d}{dt}\eta\big(t,x_0(t),u(t)\big)\lambda_j^*(t)g_{x'}^j\big(t,u,\dot{u}\big)\Big]dt$$

$$=-\int_a^b \sum_{j=1}^m \eta\big(t,x_0(t),u(t)\big)\tau_i f_x^j\big(t,u,\dot{u}\big)$$

$$+\frac{d}{dt}\eta\big(t,x_0(t),u(t)\big)\tau_i f_x^j\big(t,u,\dot{u}\big)dt$$

From (6.12) and (6.10), we have

$$\int_a^b \sum_{j=1}^m \Big[\eta\big(t,x_0(t),u(t)\big)\lambda_j^*(t)g_x^j\big(t,u,\dot{u}\big) \qquad (6.13)$$

$$+\frac{d}{dt}\eta\big(t,x_0(t),u(t)\big)\lambda_j^*(t)g_{x'}^j\big(t,u,\dot{u}\big)\Big]dt > 0$$

Now (6.13) in view of $V-$quasi-invexity of

$$\left(\int_a^b \lambda_1 g_1(t,\cdot,\cdot)dt,...,\int_a^b \tau_m g_m(t,\cdot,\cdot)dt\right)$$

yields

$$\int_a^b \sum_{j=1}^m \beta_j\left(t,x_0(t),u(t)\right)\lambda_j^* g_j\left(t,x_0,\dot{x}_0\right)dt$$

$$> \int_a^b \sum_{j=1}^m \beta_j\left(t,x_0(t),u(t)\right)\lambda_j^* g_j\left(t,u,\dot{u}\right)dt$$

This is a contradiction, since $\beta_j\left(t,\ x_0(t),u(t)\right)\lambda_j^*(t)g_j\left(t,x_0,\dot{x}_0\right)\le 0,$

$$\beta_j\left(t,\ x_0(t),u(t)\right)\lambda_j^*(t)g_j\left(t,u,\dot{u}\right)= 0,$$

and

$$\beta_j\left(t,\ x_0(t),u(t)\right)> 0,\quad j=1,...,m\ .$$

6.4 Mond-Weir type Duality

In this Section we consider the dual (VCD) given in Section 2 of this Chapter and establish duality results under generalized $V-$ invexity assumption on the functions involved.

Theorem 6.4.1 (Weak Duality): Let x be feasible for (VCP) and let $(u,\ \tau,\ \lambda)$ be feasible for (VCD). If $\left(\int_a^b \tau_1 f_1(t,\cdot,\cdot)dt,...,\int_a^b \tau_p f_p(t,\cdot,\cdot)dt\right)$

is $V-$ pseudo-invex and $\left(\int_a^b \lambda_1 g_1(t,\cdot,\cdot)dt,...,\int_a^b \tau_m g_m(t,\cdot,\cdot)dt\right)$ is

$V-$ quasi-invex with respect to η, then

$$\left(\int_a^b f_1(t,x,\dot{x})dt,...,\int_a^b f_p(t,x,\dot{x})dt\right)^T$$

$$-\left(\int_a^b f_1(t,u,\dot{u})dt,...,\int_a^b f_p(t,u,\dot{u})dt\right)^T \notin -\operatorname{int} R_+^p.$$

Proof: From the feasibility conditions,

$$\int_a^b \lambda_j g_j(t,x,\dot{x})dt \le \int_a^b \lambda_j g_j(t,u,\dot{u})dt,\quad j=1,...,m.$$

Since $\beta_j(t,x,u)>0,\quad j=1,...,m,$ we have

$$\int_a^b \sum_{j=1}^m \lambda_j \beta_j (t,x,u) g_j (t,x,\dot{x}) dt \le \int_a^b \sum_{j=1}^m \lambda_j \beta_j (t,x,u) g_j (t,u,\dot{u}) dt . \quad (6.14)$$

Hence,

$$\int_a^b \sum_{j=1}^m \Big[\eta(t,x(t),u(t)) \lambda_j (t) g_u^j (t,u,\dot{u}) \quad (6.15)$$

$$+ \frac{d}{dt} \eta(t,x(t),u(t)) \lambda_j (t) g_{\dot{u}}^j (t,u,\dot{u}) \Big] dt \le 0$$

The constraint (6.4), as earlier, is equivalent to

$$\int_a^b \sum_{j=1}^m \Big[\eta(t,x(t),u(t)) \lambda_j (t) g_u^j (t,u,\dot{u}) \quad (6.16)$$

$$+ \frac{d}{dt} \eta(t,x(t),u(t)) \lambda_j (t) g_{\dot{u}}^j (t,u,\dot{u}) \Big] dt$$

$$= -\int_a^b \sum_{i=1}^p \eta(t,x(t),u(t)) \tau_i f_u^i (t,u,\dot{u})$$

$$+ \frac{d}{dt} \eta(t,x(t),u(t)) \tau_i f_{\dot{u}}^i (t,u,\dot{u}) \Big] dt$$

From (6.15) and (6.16), we get

$$\int_a^b \sum_{i=1}^p \eta(t,x(t),u(t)) \tau_i f_u^i (t,u,\dot{u}) + \frac{d}{dt} \eta(t,x(t),u(t)) \tau_i f_{\dot{u}}^i (t,u,\dot{u}) \, dt \ge 0. \quad (6.17)$$

The conclusion now follows from the V − pseudo-invexity of

$$\left(\int_a^b \tau_1 f_1 (t,\cdot,\cdot) dt , ..., \int_a^b \tau_p f_p (t,\cdot,\cdot) dt \right)$$

and $\alpha_i (t, x_0 (t), u(t)) > 0, \quad i = 1,...,p, \quad \tau e = 1$.

Theorem 6.4.2 (Strong Duality): Assume that u is a weak minimum for (VCP) and that a suitable constraint qualification is satisfied at u. Then there exist (τ, λ) such that (u, τ, λ) is feasible for (VCD) and the objective functions of (VCP) and (VCD) are equal at these points. If, also for all feasible (u, τ, λ),

$$\left(\int_a^b \tau_1 f_1 (t,\cdot,\cdot) dt , ..., \int_a^b \tau_p f_p (t,\cdot,\cdot) dt \right)$$

is V – pseudo-invex and

$$\left(\int\limits_a^b \lambda_1 g_1(t,\cdot,\cdot) dt,\ldots, \int\limits_a^b \tau_m g_m(t,\cdot,\cdot) dt \right)$$

is V – quasi-invex, then (u, τ, λ) is weak maximum for (VCD).

Proof: Since u is a weak minimum for (VCP) and a constraint qualification is satisfied at u, from the Lagrangian conditions (Theorem 6.3.1), there exists (τ, λ) such that (u, τ, λ) is feasible for (VCD). Clearly the values of (VCP) and (VCD) are equal at u, since the objective functions for both problems are the same. By the generalized V – invexity hypothesis, weak duality holds; hence if (u, τ, λ) is not a weak optimum for (VCD), there must exist (x, τ^*, λ^*) feasible for (VCD), such that

$$\left(\int\limits_a^b f_1(t,x,\dot{x}) dt,\ldots, \int\limits_a^b f_p(t,x,\dot{x}) dt \right)^T$$

$$- \left(\int\limits_a^b f_1(t,u,\dot{u}) dt,\ldots, \int\limits_a^b f_p(t,u,\dot{u}) dt \right)^T \in -\text{int } R_+^p.$$

contradicting weak duality.

The results of present Section are extended to control problem in the next Section.

6.5 Duality for Multiobjective Control Problems

A number of duality theorems for single objective control problem have been appeared in the literature (see, e.g., Hanson (1964), Kreindler (1966), Pearson (1965), Ringlee (1965), Mond and Hanson (1968), and Mond and Smart (1988)). In general, these references give conditions under which an external solution of the control problem yields a solution of the corresponding dual. Mond and Hanson (1968) established the converse duality theorem which gives conditions under which a solution of the dual problem yields a solution to the control problem. Mond and Smart (1988) extended the results of Mond and Hanson (1964) for duality in control problems to invex functions. Bhatia and Kumar (1995) extended the work of Mond and Hanson (1968) to the content of multiobjective control problems and established duality results for Wolfe as well as Mond-Weir type duals under ρ – invexity assumptions and its generalizations. The reader is refer to Kim et al. (1993) also.

In this Section we will obtain duality results for multiobjective control problems under V − invexity assumptions and its generalizations.

The control problem is to transfer the state variable from an initial state $x(a) = \alpha$ at $t = a$ to a final state $x(b) = \beta$ at $t = b$ so as to minimize a given functional, subject to constraints on the control and state variables, that is: (VCP)

$$\text{Min} \int_a^b f\left(t, x(t), u(t)\right) dt = \left(\int_a^b f_1\left(t, x(t), u(t)\right) dt, \dots, \int_a^b f_p\left(t, x(t), u(t)\right) dt \right)$$

$$\text{subject to} \qquad x(a) = \alpha, \; x(b) = \beta, \tag{6.18}$$

$$g\left(t, x(t), u(t)\right) \le 0, \quad t \in I, \tag{6.19}$$

$$h\left(t, x(t), u(t)\right) = \dot{x}, \; t \in I. \tag{6.20}$$

$x(t)$ and $u(t)$ are required to be piecewise smooth functions on I; their derivatives are continuous except perhaps at points of discontinuity of $u(t)$, which has piecewise continuous first and second derivatives.

Throughout this Section, R^n denotes an $n-$ dimensional euclidan space. , Each $f_i : I \times R^n \times R^m \rightarrow R$ for

$$i = 1, \dots, p, \quad g : I \times R^n \times R^m \rightarrow R^k$$

and for $h : I \times R^n \times R^m \rightarrow R^q$ are continuously differentiable functions. Let $x : I \rightarrow R^n$ be differentiable with its derivative $\dot{x}$ and let $y : I \rightarrow R^m$ be a smooth function. Denote the first partial derivatives of f_i with respect to $t, x, \dot{x}, y$ and z, by $f_{it}, f_{ix}, f_{i\dot{x}}, f_{iy}$ and f_{iz}, respectively; i.e.

$$f_{it} = \frac{\partial f_i}{\partial t}, f_{ix} = \left(\frac{\partial f_i}{\partial x_1}, \dots, \frac{\partial f_i}{\partial x_n} \right)^T,$$

$$f_{i\dot{x}} = \left(\frac{\partial f_i}{\partial \dot{x}_1}, \dots, \frac{\partial f_i}{\partial \dot{x}_n} \right)^T, f_{iy} = \left(\frac{\partial f_i}{\partial y_1}, \dots, \frac{\partial f_i}{\partial y_n} \right)^T f_{iz} = \left(\frac{\partial f_i}{\partial z_1}, \dots, \frac{\partial f_i}{\partial z_n} \right)^T$$

$i = 1, 2, \dots, p,$ where T denotes the transpose operator. The partial derivatives of the vector functions g and h are defined similarly, using $n \times q$ matrix and $n \times n$ matrix, respectively.

For an $r-$ dimensional vector function $R(t, x(t), \dot{x}(t), y(t), z(t))$, we denote the first partial derivative with respect to $t, x(t), \dot{x}(t), y(t)$ and $z(t)$ by $R_{i_t}, R_{i_x}, R_{i_{\dot{x}}}, R_{i_y}$ and R_{i_z}, respect $R_t = \left(\dfrac{\partial R^1}{\partial t}, \dfrac{\partial R^2}{\partial t}, \ldots, \dfrac{\partial R^r}{\partial t} \right)$,

$$R_x = \begin{pmatrix} \dfrac{\partial R^1}{\partial x_1}, \dfrac{\partial R^1}{\partial x_2}, \ldots, \dfrac{\partial R^1}{\partial x_n} \\ \dfrac{\partial R^2}{\partial x_1}, \dfrac{\partial R^2}{\partial x_2}, \ldots, \dfrac{\partial R^2}{\partial x_n} \\ \cdots \\ \cdots \\ \cdots \\ \dfrac{\partial R^r}{\partial x_1}, \dfrac{\partial R^r}{\partial x_2}, \ldots, \dfrac{\partial R^r}{\partial x_n} \end{pmatrix}_{r \times n}, R_{\dot{x}} = \begin{pmatrix} \dfrac{\partial R^1}{\partial \dot{x}_1}, \dfrac{\partial R^1}{\partial \dot{x}_2}, \ldots, \dfrac{\partial R^1}{\partial \dot{x}_n} \\ \dfrac{\partial R^2}{\partial \dot{x}_1}, \dfrac{\partial R^2}{\partial \dot{x}_2}, \ldots, \dfrac{\partial R^2}{\partial \dot{x}_n} \\ \cdots \\ \cdots \\ \cdots \\ \dfrac{\partial R^r}{\partial \dot{x}_1}, \dfrac{\partial R^r}{\partial \dot{x}_2}, \ldots, \dfrac{\partial R^r}{\partial \dot{x}_n} \end{pmatrix}_{r \times n}$$

$$R_y = \begin{pmatrix} \dfrac{\partial R^1}{\partial y_1}, \dfrac{\partial R^1}{\partial y_2}, \ldots, \dfrac{\partial R^1}{\partial y_n} \\ \dfrac{\partial R^2}{\partial y_1}, \dfrac{\partial R^2}{\partial y_2}, \ldots, \dfrac{\partial R^2}{\partial y_n} \\ \cdots \\ \cdots \\ \cdots \\ \dfrac{\partial R^r}{\partial y_1}, \dfrac{\partial R^r}{\partial y_2}, \ldots, \dfrac{\partial R^r}{\partial y_n} \end{pmatrix}_{r \times n}, R_z = \begin{pmatrix} \dfrac{\partial R^1}{\partial z_1}, \dfrac{\partial R^1}{\partial z_2}, \ldots, \dfrac{\partial R^1}{\partial z_n} \\ \dfrac{\partial R^2}{\partial z_1}, \dfrac{\partial R^2}{\partial z_2}, \ldots, \dfrac{\partial R^2}{\partial z_n} \\ \cdots \\ \cdots \\ \cdots \\ \dfrac{\partial R^r}{\partial z_1}, \dfrac{\partial R^r}{\partial z_2}, \ldots, \dfrac{\partial R^r}{\partial z_n} \end{pmatrix}_{r \times n}$$

Denote by X the space of piecewise smooth control functions $x : I \to R^n$ with norm $\|x\|_\infty$; by Z the space of piecewise continuous control functions $z : I \to R^m$ with norm $\|z\|_\infty$; by Y the space of piecewise continuous differentiable state functions $y : I \to R^n$ with norm $\|y\| = \|y\|_\infty + \|Dy\|_\infty$, where the differentiation operator D is given by

$$u = Dx \Leftrightarrow x(t) = u(a) + \int_a^b u(s)ds,$$

where $u(a)$ is a given boundary value. Therefore $\dfrac{d}{dt} = D$ except at discontinuities.

Define $\Lambda^+ = \left\{ \tau \in R^p : \tau > 0, \tau^T e = 1, e = (1,1,...,1)^T \in R^p \right\}$. Let R_+^p be the non-negative orthant of R^p.

Mond-Weir type dual for (VCP) is proposed and duality relationships are established under generalized $V - $ invexity assumptions:

(MVCD)   Maximize $\left(\int_a^b f_1(t, y(t), v(t)) dt, \ ..., \ \int_a^b f_p(t, y(t), v(t)) dt \right)$

subject to $y(a) = v(a) = 0, \ y(b) = v(b) = 0,$ (6.21)

$$\sum_{i=1}^p \tau_i f_{iy}(t, y(t), v(t)) + \sum_{j=1}^m \lambda_j(t) g_{jy}(t, y(t), v(t)) \quad (6.22)$$

$$+ \sum_{r=1}^q \mu_r(t) h_{ry}(t, y(t), v(t)) + \dot{u}(t) = 0, \quad t \in I,$$

$$D\left[\sum_{i=1}^p \lambda_i f_{ix}(t, u(t), \dot{u}(t), v(t), w(t)) \right. \quad (6.23)$$

$$+ \mu(t) g_{\dot{x}}(t, u(t), \dot{u}(t), v(t), w(t))$$

$$\left. \rho(t)^T h_{\dot{x}}(t, u(t), \dot{u}(t), v(t), w(t)) \right] = 0, t \in I$$

$$\sum_{i=1}^p \tau_i f_{iy}(t, y(t), v(t)) + \sum_{j=1}^m \lambda_j(t) g_{jv}(t, y(t), v(t)) \quad (6.24)$$

$$+ \sum_{r=1}^q \mu_r(t) h_{rv}(t, y(t), v(t)) = 0, t \in I$$

$$\int_a^b \sum_{j=1}^m \lambda_j(t) g_j(t, y(t), v(t)) dt \geq 0, \quad t \in I,$$

$$\int_a^b \sum_{r=1}^q \mu_r(t) [h(t, y(t), v(t)) - \dot{x}(t)] dt \geq 0, \quad t \in I, \quad (6.25)$$

$$\lambda(t) \geq 0, t \in I, \tau_i \geq 0, i = 1, \ldots, p, \sum_{i=1}^{p} \tau_i = 1 \qquad (6.26)$$

Optimization in (VCP) and (MVCD) means obtaining efficient solutions for the corresponding programs.

Let $\;F_i(x) = \int_a^b f_i(t, x, u)\,dt\;$, $\quad i = 1, \ldots, p\;$ be Frechet differentiable.

Let there exist functions

$$v\left(t, x, \overline{x}, \dot{x}, \dot{\overline{x}}, u, \overline{u}\right) \in R^p$$

and

$$\eta\left(t, x, x^*, \dot{x}, \dot{x}^*, u, u^*\right) \in R^n$$

with $\eta = 0$ at t if $x(t) = \overline{x}(t)$ and $\xi\left(t, x, \overline{x}, \dot{x}, \dot{\overline{x}}, u, \overline{u}\right) \in R^m$.

Definition 6.5.1 ($V - $ Invex): A vector function $F = \left(F_1, \ldots, F_p\right)$ is said to be $V - $ invex in $x, \overline{x}$ and u with respect to η, ξ and α if there exists differentiable vector function $\eta : I \times R^n \times R^n \to R^n$ with $\eta(t, x, x) = 0$, $\xi\left(t, x, \overline{x}, \dot{x}, \dot{\overline{x}}, u, \overline{u}\right) \in R^m$ and $\alpha_i : I \times X \times X \to R_+ \setminus \{0\}$ such that for each $x, \overline{x} \in X$ and $u, \overline{u} \in Y$ for $i = 1, \ldots, p$

$$F_i(x) - F_i(\overline{x}) \geq \int_a^b \Big[\alpha_i(t, x, \overline{x}, \dot{x}, \dot{\overline{x}}, u, \overline{u}) f_x^i(t, x, \dot{\overline{x}}, \overline{u}) \eta(t, x, \overline{x}, \dot{x}, \dot{\overline{x}}, u, \overline{u})$$

$$+ \frac{d}{dt} \eta(t, x, \overline{x}, \dot{x}, \dot{\overline{x}}, u, \overline{u}) \alpha_i(t, x, \overline{x}, \dot{x}, \dot{\overline{x}}, u, \overline{u}) f_{\overline{x}}^i(t, \overline{x}, \dot{\overline{x}}, \overline{u})$$

$$+ \alpha_i(t, x, \overline{x}, \dot{x}, \dot{\overline{x}}, u, \overline{u}) h_u^i(t, \overline{x}, \dot{\overline{x}}, \overline{u}) \xi(t, x, \overline{x}, \dot{x}, \dot{\overline{x}}, u, \overline{u}) \Big] dt$$

Definition 6.5.2 ($V - $ Pseudo-Invex):

A vector function $F = \left(F_1, \ldots, F_p\right)$ is said to be $V - $ pseudo-invex in $x, \overline{x}$ and u with respect to η, ξ and β if there exists differentiable vector function $\eta : I \times R^n \times R^n \to R^n$ with $\eta(t, x, x) = 0$,

$$\xi\left(t, x, \overline{x}, \dot{x}, \dot{\overline{x}}, u, \overline{u}\right) \in R^m \text{ and } \beta_i : I \times X \times X \to R_+ \setminus \{0\}$$

such that for each $x, \overline{x} \in X$ and $u, \overline{u} \in Y$ for $i = 1, \ldots, p$

$$\int_a^b \sum_{i=1}^p \left[\eta(t,x,\overline{x},\dot{x},\dot{\overline{x}},u,\overline{u}) f_x^i(t,x,\dot{\overline{x}},\overline{u}) \right.$$

$$+ \frac{d}{dt}\eta(t,x,\overline{x},\dot{x},\dot{\overline{x}},u,\overline{u}) f_{\dot{\overline{x}}}^i(t,\overline{x},\dot{\overline{x}},\overline{u})$$

$$\left. + h_u^i(t,\overline{x},\dot{\overline{x}},\overline{u}) \xi(t,x,\overline{x},\dot{x},\dot{\overline{x}},u,\overline{u}) \right] dt \geq 0$$

$$\Rightarrow \int_a^b \sum_{i=1}^p \beta_i(t,x,\overline{x},\dot{x},\dot{\overline{x}},u,\overline{u}) f_i(t,x,\dot{x},u) dt$$

$$\geq \int_a^b \sum_{i=1}^p \beta_i(t,x,\overline{x},\dot{x},\dot{\overline{x}},u,\overline{u}) f_i(t,\overline{x},\dot{\overline{x}},\overline{u}) dt,$$

Or equivalently;

$$\int_a^b \sum_{i=1}^p \beta_i(t,x,\overline{x},\dot{x},\dot{\overline{x}},u,\overline{u}) f_i(t,x,\dot{x},u) dt$$

$$< \int_a^b \sum_{i=1}^p \beta_i(t,x,\overline{x},\dot{x},\dot{\overline{x}},u,\overline{u}) f_i(t,\overline{x},\dot{\overline{x}},\overline{u}) dt$$

$$\int_a^b \sum_{i=1}^p \left[\eta(t,x,\overline{x},\dot{x},\dot{\overline{x}},u,\overline{u}) f_x^i(t,x,\dot{\overline{x}},\overline{u}) \right.$$

$$+ \frac{d}{dt}\eta(t,x,\overline{x},\dot{x},\dot{\overline{x}},u,\overline{u}) f_{\dot{\overline{x}}}^i(t,\overline{x},\dot{\overline{x}},\overline{u})$$

$$\left. + h_u^i(t,\overline{x},\dot{\overline{x}},\overline{u}) \xi(t,x,\overline{x},\dot{x},\dot{\overline{x}},u,\overline{u}) \right] dt < 0$$

Definition 6.5.3 $(V - \textbf{Quasi-Invex})$:

A vector function $F = (F_1,...,F_p)$ is said to be $V-$ quasi-invex in $x,\overline{x}$ and u with respect to $\eta,\ \xi$ and β if there exists differentiable vector function $\eta : I \times R^n \times R^n \to R^n$ with $\eta(t,x,x) = 0$,

$$\xi(t,x,\overline{x},\dot{x},\dot{\overline{x}},u,\overline{u}) \in R^m \text{ and } \beta_i : I \times X \times X \to R_+ \setminus \{0\}$$

such that for each $x,\ \overline{x} \in X$ and $u,\ \overline{u} \in Y$ for $i = 1,...,p$

$$\int_a^b \sum_{i=1}^p \beta_i(t,x,\overline{x},\dot{x},\dot{\overline{x}},u,\overline{u})f_i(t,x,\dot{x},u)dt$$

$$\leq \int_a^b \sum_{i=1}^p \beta_i(t,x,\overline{x},\dot{x},\dot{\overline{x}},u,\overline{u})f_i(t,\overline{x},\dot{\overline{x}},\overline{u})dt$$

$$\Rightarrow \int_a^b \sum_{i=1}^p \Big[\eta(t,x,\overline{x},\dot{x},\dot{\overline{x}},u,\overline{u})f_x^i(t,x,\dot{\overline{x}},\overline{u})$$

$$+\frac{d}{dt}\eta(t,x,\overline{x},\dot{x},\dot{\overline{x}},u,\overline{u})f_{\overline{x}}^i(t,\overline{x},\dot{\overline{x}},\overline{u})$$

$$+h_u^i(t,\overline{x},\dot{\overline{x}},\overline{u})\xi(t,x,\overline{x},\dot{x},\dot{\overline{x}},u,\overline{u})\Big]dt \leq 0$$

Or equivalently;

$$\int_a^b \sum_{i=1}^p \Big[\eta(t,x,\overline{x},\dot{x},\dot{\overline{x}},u,\overline{u})f_x^i(t,x,\dot{\overline{x}},\overline{u})$$

$$+\frac{d}{dt}\eta(t,x,\overline{x},\dot{x},\dot{\overline{x}},u,\overline{u})f_{\overline{x}}^i(t,\overline{x},\dot{\overline{x}},\overline{u})$$

$$+h_u^i(t,\overline{x},\dot{\overline{x}},\overline{u})\xi(t,x,\overline{x},\dot{x},\dot{\overline{x}},u,\overline{u})\Big]dt > 0$$

$$\Rightarrow \int_a^b \sum_{i=1}^p \beta_i(t,x,\overline{x},\dot{x},\dot{\overline{x}},u,\overline{u})f_i(t,x,\dot{x},u)dt$$

$$> \int_a^b \sum_{i=1}^p \beta_i(t,x,\overline{x},\dot{x},\dot{\overline{x}},u,\overline{u})f_i(t,\overline{x},\dot{\overline{x}},\overline{u})dt,$$

Remark 6.5.1: $V -$ invexity is defined here for functionals instead of functions, unlike the definition given in Section 1. This has been done so that $V -$ invexity of a functional is necessary and sufficient for its critical points to be global minima, which coincide with the original concept of a $V -$ invex function being one for which critical points are also global minima (Craven and Glover (1985)).

We thus have the following characterization result.

Lemma 6.5.1: $F(x) = \int_a^b f(t,x,\dot{x},u)dt$ is $V -$ invex if and only if every critical point F is global minimum.

Note 6.5.1: $(\bar{x}(t),\,\bar{u}(t))$ is a critical point of F if

$$f_x^i(t,\bar{x},\dot{\bar{x}},\bar{u}) = \frac{d}{dt}f_{\dot{x}}^i(t,\bar{x},\dot{\bar{x}},\bar{u}) \text{ and } f_u^i(t,\bar{x},\dot{\bar{x}},\bar{u}) = 0$$

almost everywhere in the interval $[a,\ b]$. If $x(a)$ and $x(b)$ are free, the transversality conditions $h_{\dot{x}}(t,\bar{x},\dot{\bar{x}},\bar{u}) = 0$ at a and b are included.

Proof of Lemma 6.5.1:

($\Rightarrow$) Assume that there exist functions $\eta,\ \xi$ and α such that F is V – invex with respect to $\eta,\ \xi$ and α on $[a,\ b]$. Let $(\bar{x},\bar{u})$ be a critical point of F. Then, for $i = 1,...,p$

$$F_i(x) - F_i(\bar{x}) \geq \int_a^b \Big[\alpha_i(t,x,\bar{x},\dot{x},\dot{\bar{x}},u,\bar{u})f_x^i(t,x,\bar{x},\bar{u})\eta(t,x,\bar{x},\dot{x},\dot{\bar{x}},u,\bar{u})$$

$$+ \frac{d}{dt}\eta(t,x,\bar{x},\dot{x},\dot{\bar{x}},u,\bar{u})\alpha_i(t,x,\bar{x},\dot{x},\dot{\bar{x}},u,\bar{u})f_{\bar{x}}^i(t,\bar{x},\dot{\bar{x}},\bar{u})$$

$$+ \alpha_i(t,x,\bar{x},\dot{x},\dot{\bar{x}},u,\bar{u})h_u^i(t,\bar{x},\dot{\bar{x}},\bar{u})\xi(t,x,\bar{x},\dot{x},\dot{\bar{x}},u,\bar{u}) \Big] dt$$

$$= \int_a^b \Big[\alpha_i(t,x,\bar{x},\dot{x},\dot{\bar{x}},u,\bar{u})f_x^i(t,x,\bar{x},\bar{u})\eta(t,x,\bar{x},\dot{x},\dot{\bar{x}},u,\bar{u})$$

$$- \eta(t,x,\bar{x},\dot{x},\dot{\bar{x}},u,\bar{u})\alpha_i(t,x,\bar{x},\dot{x},\dot{\bar{x}},u,\bar{u})\frac{d}{dt}f_{\dot{x}}^i(t,x,\bar{x},\bar{u})$$

$$+ \alpha_i(t,x,\bar{x},\dot{x},\dot{\bar{x}},u,\bar{u})f_u^i(t,x,\bar{x},\bar{u})\xi(t,x,\bar{x},\dot{x},\dot{\bar{x}},u,\bar{u}) \Big] dt$$

$$+ \eta(t,x,\bar{x},\dot{x},\dot{\bar{x}},u,\bar{u})f_{\bar{x}}^i(t,x,\dot{\bar{x}},\bar{u})$$

$$= 0$$

as $(\bar{x},\bar{u})$ is a critical point of either fixed boundary conditions imply that $\eta = 0$ at a and b or free boundary conditions imply that $f_{\dot{x}}^{'} = 0$ at a and b. Therefore, $(\bar{x},\bar{u})$ is a global minimum of F.

($\Leftarrow$) Assume that every critical point is a global minimum.

If $(\bar{x},\bar{u})$ is a critical point, then if $f_x^i \neq \frac{d}{dt}f_{\dot{x}}^i$ at $(\bar{x},\bar{u})$, put

$$\eta_i = \frac{f^i(t,x,\dot{x},u) - f^i(t,\bar{x},\dot{\bar{x}},\bar{u})}{2\left(f_x^i - \frac{d}{dt}f_{\dot{x}}^i\right)^T \left(f_x^i - \frac{d}{dt}f_{\dot{x}}^i\right)}\left(f_x^i - \frac{d}{dt}f_{\dot{x}}^i\right)$$

$\alpha = 1$, or if $f_{\dot{x}}^i = \dfrac{d}{dt} f_{\dot{x}}^i$, put $\eta = 0$; and if $h_u \neq 0$, put

$$\xi_i = \frac{f^i(t,x,\dot{x},u) - f^i(t,\overline{x},\dot{\overline{x}},\overline{u})}{2(f_u^i)^T (f_u^i)} (f_u^i)$$

and $\alpha = 1$, or if $f_u^i = 0$, put $\xi = 0$.

Mond and Hanson (1968) pointed out that if the primal solution for (VCP) is normal, then Fritz-John conditions reduce to Kuhn-Tucker conditions.

Lemma 6.5.2. (Kuhn-Tucker Necessary Optimality Conditions):

If $(\overline{x},\overline{u}) \in X \times Y$ solves (VCP) if the Frechet derivative $D^- F_x^i (x^0, u^0)$ is surjective, and if the optimal solution (x^0, u^0) is normal, then there exist piecewise smooth $\tau^0 : I \to R^p$, $\lambda^0 : I \to R^m$ and $\mu^0 : I \to R^k$, satisfying the following for all $t \in [a,\ b]$:

$$\sum_{i=1}^{p} \tau_i^0 f_x^i(t,x^0,u^0) + \sum_{j=1}^{m} \lambda_j^0(t) g_x^j(t,x^0,u^0) \tag{6.27}$$

$$+ \sum_{r=1}^{q} \mu_r^0(t) h_x^r(t,x^0,u^0) + \mu_r^0(t) = 0, \ \ t \in I,$$

$$\sum_{i=1}^{p} \tau_i^0 f_x^i(t,x^0,u^0) + \sum_{j=1}^{m} \lambda_j^0(t) g_x^j(t,x^0,u^0) \tag{6.28}$$

$$+ \sum_{r=1}^{q} \mu_r^0(t) h_x^r(t,x^0,u^0) = 0, \ \ t \in I,$$

$$\sum_{j=1}^{m} \lambda_j^0(t) g_j(t,x^0,u^0) = 0, \ \ t \in I, \tag{6.29}$$

$$\lambda^0(t) \geq 0, \ \ t \in I, \ \ \tau_i^0 > 0, \ \ i = 1,...,p, \ \ \sum_{i=1}^{p} \tau_i^0 = 1. \tag{6.30}$$

We shall now prove that (VCP) and MVCD) are a dual pair subject to generalized V − invexity conditions on the objective and constraint functions.

Theorem 6.5.1 (Weak Duality): Assume that for all feasible $(x,\ u)$ for (VCP) and all feasible $(y,\ v,\tau,\lambda,\mu)$ for (MVCD). If

$$\left(\int_a^b \tau_1 f_1(t,\cdot,\cdot,\cdot)\, dt,\ldots, \int_a^b \tau_p f_p(t,\cdot,\cdot,\cdot)\, dt \right)$$

and

$$\left(\int_a^b \lambda_1 g_1(t,\cdot,\cdot,\cdot)\, dt,\ldots, \int_a^b \tau_m g_m(t,\cdot,\cdot,\cdot)\, dt \right)$$

are V – quasi-invex and

$$\left(\int_a^b \mu_1 \big[h_1(t,\cdot,\cdot,\cdot) - \dot{x}\big]\, dt,\ldots, \int_a^b \mu_k \big[h_k(t,\cdot,\cdot,\cdot) - \dot{x}\big]\, dt \right)$$

is strictly V – quasi-invex. Then the following can not hold:

$$\int_a^b f_i(t,x,u)\, dt \le \int_a^b f_i(t,y,v)\, dt, \quad \forall\, i = 1,\ldots,p \tag{6.31}$$

$$\int_a^b f_{i_0}(t,x,u)\, dt < \int_a^b f_{i_0}(t,y,v)\, dt, \quad \text{for some } i_0 \in \{1,\ldots,p\}. \tag{6.32}$$

Proof: Suppose contrary to the result that (6.31) and (6.32) hold. Then by V – quasi-invexity, we get

$$\int_a^b \sum_{i=1}^p \big[\eta(t,x,\overline{x},\dot{x},\dot{\overline{x}},u,\overline{u})\tau_i f_y^i(t,y,\dot{y},v) \tag{6.33}$$

$$+ f_v^i(t,\overline{x},\dot{\overline{x}},\overline{u})\xi(t,x,\overline{x},\dot{x},\dot{\overline{x}},u,\overline{u}) \big]\, dt < 0$$

From the feasibility conditions,

$$\int_a^b \lambda_j g_j(t,x,\dot{x},u)\, dt \le \int_a^b \lambda_j g_j(t,\overline{x},\dot{\overline{x}},\overline{u})\, dt, \quad \forall\, j = 1,\ldots,m.$$

Since $\beta_j(t,x,\overline{x},\dot{x},\dot{\overline{x}},u,\overline{u}) > 0, \quad \forall\, j = 1,\ldots,m$, we have

$$\int_a^b \sum_{j=1}^m \beta_j(t,x,\overline{x},\dot{x},\dot{\overline{x}},u,\overline{u})\lambda_j g_j(t,x,\dot{x},u)\, dt$$

$$\le \int_a^b \sum_{j=1}^m \beta_j(t,x,\overline{x},\dot{x},\dot{\overline{x}},u,\overline{u})\lambda_j g_j(t,\overline{x},\dot{\overline{x}},\overline{u})\, dt$$

Then by V – quasi-invexity of

$$\left(\int_a^b \lambda_1 g_1(t,\cdot,\cdot,\cdot)\, dt,\ldots, \int_a^b \tau_m g_m(t,\cdot,\cdot,\cdot)\, dt \right),$$

Gives

$$\int_a^b \sum_{j=1}^m \left[\eta(t,x,\overline{x},\dot{x},\dot{\overline{x}},u,\overline{u}) \lambda_j g_x^j(t,\overline{x},\dot{\overline{x}},\overline{u}) \right. \tag{6.34}$$

$$\left. + g_u^j(t,\overline{x},\dot{\overline{x}},\overline{u})\xi(t,x,\overline{x},\dot{x},\dot{\overline{x}},u,\overline{u}) \right] dt \le 0$$

Similarly, we have

$$\int_a^b \sum_{k=1}^q \gamma_k(t,x,\overline{x},\dot{x},\dot{\overline{x}},u,\overline{u})\mu_k \left[h_k(t,x,\dot{x},u) - \dot{x} \right] dt$$

$$\le \int_a^b \sum_{k=1}^q \gamma_k(t,x,\overline{x},\dot{x},\dot{\overline{x}},u,\overline{u})\mu_k \left[h_k(t,\overline{x},\dot{\overline{x}},\overline{u}) - \dot{x} \right] dt$$

From strict $V - $quasi-invexity of

$$\left(\int_a^b \mu_1 [h_1(t,\cdot,\cdot,\cdot) - \dot{x}] dt, \ldots, \int_a^b \mu_k [h_k(t,\cdot,\cdot,\cdot) - \dot{x}] dt \right),$$

we have

$$\int_a^b \sum_{k=1}^q \left[\eta(t,x,\overline{x},\dot{x},\dot{\overline{x}},u,\overline{u}) \, \mu_k h_y^k(t,y,\dot{y},v) - \frac{d}{dt}\eta(t,x,\overline{x},\dot{x},\dot{\overline{x}},u,\overline{u})\mu_k \right. \tag{6.35}$$

$$\left. + \sum_{k=1}^q \mu_k h_v^k(t,y,\dot{y},v)\xi(t,x,\overline{x},\dot{x},\dot{\overline{x}},u,\overline{u}) \right] dt < 0$$

By integrating $\dfrac{d}{dt}\eta(t,x,\overline{x},\dot{x},\dot{\overline{x}},u,\overline{u})\mu_k$ from a to b by parts and applying the boundary conditions (6.18), we have

$$\int_a^b \frac{d}{dt}\eta(t,x,\overline{x},\dot{x},\dot{\overline{x}},u,\overline{u})\mu_k dt = -\int_a^b \eta(t,x,\overline{x},\dot{x},\dot{\overline{x}},u,\overline{u})\mu_k \, dt . \tag{6.36}$$

Using (6.36) in (6.35), we have

$$\int_a^b \sum_{k=1}^q \left[\eta(t,x,\overline{x},\dot{x},\dot{\overline{x}},u,\overline{u}) \, \mu_k h_y^k(t,y,\dot{y},v) + \mu^0 \right. \tag{6.37}$$

$$\left. + \sum_{k=1}^q \mu_k h_v^k(t,y,\dot{y},v)\xi(t,x,\overline{x},\dot{x},\dot{\overline{x}},u,\overline{u}) \right] dt < 0$$

From (6.33), (6.34) and (6.37), we have

$$\int_a^b \left\{ \eta \left[\sum_{i=1}^p \tau_i f_y^i(t,y,\dot{y},v) + \sum_{j=1}^m \lambda_j g_y^j(t,y,\dot{y},v) + \sum_{r=1}^k \mu_r f_y^r(t,y,\dot{y},v) \right] + \right.$$

$$\left. \xi \left[\sum_{i=1}^p \tau_i f_v^i(t,y,\dot{y},v) + \sum_{j=1}^m \lambda_j g_v^j(t,y,\dot{y},v) + \sum_{r=1}^k \mu_r f_y^r(t,y,\dot{y},v) \right] \right\} dt < 0$$

which is a contradiction to (6.22) and (6.23).

Corollary 6.5.1: Assume that weak duality (Theorem 6.5.1) holds between (VCP) and (MVCD). If (y, v) is feasible for (VCP) and $(y, v, \tau, \lambda, \mu)$ is feasible for (MVCD), then (y, v) is efficient for (VCP) and $(y, v, \tau, \lambda, \mu)$ is efficient for (MVCD).

Proof: Suppose (y, v) is not efficient for (VCP). Then there exists some feasible (x, u) for (VCP) such that

$$\int_a^b f_i(t,x,\dot{x},u)\,dt \le \int_a^b f_i(t,y,\dot{y},v)\,dt, \quad \forall \, i=1,...,p,$$

$$\int_a^b f_{i_0}(t,x,\dot{x},u)\,dt < \int_a^b f_{i_0}(t,y,\dot{y},v)\,dt, \quad \text{for some } i_0 \in \{1,...,p\}.$$

This contradicts weak duality. Hence (y, v) is efficient for (VCP). Now suppose $(y, v, \tau, \lambda, \mu)$ is not efficient for (MVCD). Then there exist some $(x, u, \tau, \lambda, \mu)$ feasible for (MVCD) such that

$$\int_a^b f_i(t,x,\dot{x},u)\,dt \le \int_a^b f_i(t,y,\dot{y},v)\,dt, \quad \forall \, i=1,...,p,$$

$$\int_a^b f_{i_0}(t,x,\dot{x},u)\,dt < \int_a^b f_{i_0}(t,y,\dot{y},v)\,dt, \quad \text{for some } i_0 \in \{1,...,p\}.$$

This contradicts weak duality. Hence $(y, v, \tau, \lambda, \mu)$ is efficient for (MVCD).

Theorem 6.5.2 (Strong Duality): Let $(\bar{x}, \bar{u})$ be efficient for (VCP) and assume that $(\bar{x}, \bar{u})$ satisfy the constraint qualification of Lemma 6.5.1 for at least one $i_0 \in \{1,...,p\}$. Then there exist $\bar{\tau} \in R^p$ and piecewise smooth $\bar{\lambda} : I \to R^m$ and $\bar{\mu} : I \to R^k$ such that $(\bar{x}, \bar{u}, \bar{\tau}, \bar{\lambda}, \bar{\mu})$ is feasible for (MVCD). If also weak duality (Theorem 6.5.1) holds between (VCP) and (MVCD) then $(\bar{x}, \bar{u}, \bar{\tau}, \bar{\lambda}, \bar{\mu})$ is efficient for (MVCD).

Proof: Proceeding on the same lines as in Theorem 6.5.1, it follows that there exist piecewise smooth $\bar{\tau} : I \to R^p$, $\bar{\lambda} : I \to R^m$ and $\bar{\mu} : I \to R^k$, satisfying for all $t \in I$ the following relations:

$$\sum_{i=1}^{p} \bar{\tau}_i f_x^i \left(t, \bar{x}, \dot{\bar{x}}, \bar{u}\right) + \sum_{j=1}^{m} \bar{\lambda}_j (t) g_x^j \left(t, \bar{x}, \dot{\bar{x}}, \bar{u}\right)$$

$$+ \sum_{r=1}^{k} \bar{\mu}_r (t) h_x^r \left(t, \bar{x}, \dot{\bar{x}}, \bar{u}\right) + \bar{u}(t) = 0, \quad t \in I,$$

$$\sum_{i=1}^{p} \bar{\tau}_i f_{\dot{x}}^i \left(t, \bar{x}, \dot{\bar{x}}, \bar{u}\right) + \sum_{j=1}^{m} \bar{\lambda}_j (t) g_{\dot{x}}^j \left(t, \bar{x}, \dot{\bar{x}}, \bar{u}\right)$$

$$+ \sum_{r=1}^{k} \bar{\mu}_r (t) h_{\dot{x}}^r \left(t, \bar{x}, \dot{\bar{x}}, \bar{u}\right) = 0, \quad t \in I,$$

$$\sum_{j=1}^{m} \bar{\lambda}_j (t) g_j \left(t, \bar{x}, \dot{\bar{x}}, \bar{u}\right) = 0, \quad t \in I,$$

$$\bar{\lambda}(t) \geq 0, \quad t \in I, \quad \bar{\tau}_i > 0, \quad i = 1, ..., p, \quad \sum_{i=1}^{p} \bar{\tau}_i = 1.$$

The relations

$$\int_a^b \sum_{j=1}^{m} \bar{\lambda}_j (t) g_j \left(t, \bar{x}, \dot{\bar{x}}, \bar{u}\right) dt \geq 0,$$

and

$$\int_a^b \sum_{r=1}^{k} \bar{\mu}_r (t) \left[h_r \left(t, \bar{x}, \dot{\bar{x}}, \bar{u}\right) - \dot{x} \right] dt \geq 0$$

are obvious.

The above relations imply that $(\bar{x}, \bar{u}, \bar{\tau}, \bar{\lambda}, \bar{\mu})$ is feasible for (MVCD). The result now follows from Corollary 6.5.1.

6.6 Duality for a Class of Nondifferentiable Multiobjective Variational Problems

In this Section, we consider a class of nondifferentiable multiobjective variational problem and establish various duality results under generalized

V – invexity assumptions on the functionals involved using the concept of conditional proper efficiency. The following definitions will be needed in the sequel:

Consider the following vector minimization problem:

$$\text{(VCP)} \quad \text{Minnimize } \int_a^b f(t,x,\dot{x})dt = \left(\int_a^b f_1(t,x,\dot{x})dt,..., \int_a^b f_p(t,x,\dot{x})dt \right)$$

$$\text{subject to} \quad x(a) = \alpha, \quad x(b) = \beta$$

$$g(t,x,\dot{x}) \le 0, \quad t \in I.$$

where $f_i : I \times R^n \times R^n \to R$, $i \in P = \{1,...,p\}$, $g : I \times R^n \times R^n \to R^m$ are assumed to be continuously differentiable functions. Let K be the set of all feasible solutions for (VCP), that is,

$$K = \{x \in X : x(a) = \alpha, x(b) = \beta, g(t, x(t), \dot{x}(t)) \le 0, t \in I\}.$$

The following Definitions will be needed in the sequel:

Definition 6.6.1: A point $x^* \in K$ is said to be an *efficient* solution for (VCP) if for all $x \in K$

$$\int_a^b f_i(t,x^*(t),\dot{x}^*(t))dt \ge \int_a^b f_i(t,x(t),\dot{x}(t))dt, \quad \text{for all } i = 1,...,p$$

$$\Rightarrow \int_a^b f_i(t,x^*(t),\dot{x}^*(t))dt = \int_a^b f_i(t,x(t),\dot{x}(t))dt, \quad \text{for all } i = 1,...,p.$$

Definition 6.6.2 [Borwein (1979)]: A point $x^* \in K$ is said to be a *weak minimum* solution for (VCP) if there exists no $x \in K$ for which

$$\int_a^b f(t,x^*(t),\dot{x}^*(t))dt > \int_a^b f(t,x(t),\dot{x}(t))dt .$$

From this it follows that if an $x^* \in K$ is efficient for (VCP) then it is a weak minimum for (VCP).

Definition 6.6.3: A point $x^* \in K$ is said to be a *properly efficient* solution for (VCP) if there exists scalar $M > 0$ such that, for all $x \in K$, for all $i = 1,...,p$,

$$\int_a^b f_i(t,x^*(t),\dot{x}^*(t))dt - \int_a^b f_i(t,x(t),\dot{x}(t))dt$$

$$\leq M\left(\int_a^b f_j(t,x(t),\dot{x}(t))\,dt - \int_a^b f_j(t,x^*(t),\dot{x}^*(t))\,dt\right)$$

for some j such that $\int_a^b f_j(t,x(t),\dot{x}(t))\,dt > \int_a^b f_j(t,x^*(t),\dot{x}^*(t))\,dt$ when-

ever $x \in K$ and $\int_a^b f_i(t,x^*(t),\dot{x}^*(t))\,dt > \int_a^b f_i(t,x(t),\dot{x}(t))\,dt$.

An efficient solution that is not properly efficient is said to be improperly efficient. Thus for x^* to be improperly efficient means that to every sufficiently large $M > 0$, there is an $x \in K$ and an index $i \in \{1,...,p\}$ such that

$$\int_a^b f_j(t,x(t),\dot{x}(t))\,dt < \int_a^b f_j(t,x^*(t),\dot{x}^*(t))\,dt$$

and

$$\int_a^b f_i(t,x^*(t),\dot{x}^*(t))\,dt - \int_a^b f_i(t,x(t),\dot{x}(t))\,dt$$

$$> M\left(\int_a^b f_j(t,x(t),\dot{x}(t))\,dt - \int_a^b f_j(t,x^*(t),\dot{x}^*(t))\,dt\right), \quad \forall\ j=1,...,p,$$

such that $\int_a^b f_i(t,x^*(t),\dot{x}^*(t))\,dt < \int_a^b f_i(t,x(t),\dot{x}(t))\,dt$.

Definition 6.6.4: A point $x^* \in K$ is said to be a *conditionally properly efficient* solution for (VCP) if there exists scalar $M(x) > 0$ such that, for all $x \in K$, for all $i=1,...,p$,

$$\int_a^b f_i(t,x^*(t),\dot{x}^*(t))\,dt - \int_a^b f_i(t,x(t),\dot{x}(t))\,dt$$

$$\leq M(x)\left(\int_a^b f_j(t,x(t),\dot{x}(t))\,dt - \int_a^b f_j(t,x^*(t),\dot{x}^*(t))\,dt\right)$$

for some j such that $\int\limits_a^b f_j(t,x(t),\dot{x}(t))dt > \int\limits_a^b f_j(t,x^*(t),\dot{x}^*(t))dt$ when-

ever $x \in K$ and $\int\limits_a^b f_i(t,x^*(t),\dot{x}^*(t))dt > \int\limits_a^b f_i(t,x(t),\dot{x}(t))dt$.

We now consider the following Singh and Hanson (1991) type parametric variational problem for predetermined positive functions $\tau_i(x)$ such that $a_i < \tau_i(x) < b_i$, $i=1,...,p$, where a_i and b_i , $i=1,...,p$ are specified constants.

(CP_τ^0) Minimize $\sum\limits_{i=1}^{p} \int\limits_a^b \tau_i(x)f^i(t,x,\dot{x})dt$

subject to $x(a) = \alpha,$ $x(b) = \beta$

$$g(t,x,\dot{x}) \leq 0, \ t \in I.$$

Problem (VCP) and (CP_τ^0) are equivalent in the sense of Singh and Hanson (1991). Theorems 6.6.1 and 6.6.2, are valid when R^n is replaced by some normed space of functions, as the proofs of these theorems do not depend on the dimensionality of the space in which the feasible set of (VCP) lies. For the variational problem in question the feasible set lies in the normed space $C(I, R^n)$. For completeness we shall merely state these theorems characterizing conditional proper efficiency of (VCP) in terms of solutions of (CP_τ^0).

Theorem 6.6.1: If x^* is an optimal solution for (CP_τ^0) then x^* is conditionally properly efficient for (CP_τ^0).

Theorem 6.6.2: If x^* is conditionally properly efficient for (VCP) then x^* is optimal for (CP_τ^0) for some $\tau_i(x^*) > 0$, $i=1,...,p$.

In the subsequent analysis, we shall frequently use the following generalized Schwarz inequality

$$x^T Bz \leq \left(x^T Bx\right)^{\frac{1}{2}}\left(z^T Bz\right)^{\frac{1}{2}},$$

where B is an $n \times n$ positive semidefinite matrix.

Consider the following nondifferentiable multiobjective variational problem:

(NVCP) $\text{Min} \int_a^b \psi(t,x,\dot{x})dt = \left(\int_a^b \left(f_1(t,x,\dot{x}) + \left(x^T(t)B^1(t)x(t)\right)\right)dt, \right.$

$$\left. \ldots, \int_a^b \left(f_1(t,x,\dot{x}) + \left(x^T(t)B^1(t)x(t)\right)\right)dt \right)$$

subject to $x(a) = \alpha, \quad x(b) = \beta$

$$g(t,x,\dot{x}) \leq 0, \quad t \in I.$$

where $B_i(t)$, $i \in P = \{1, \ldots, p\}$, is a positive semi-definite (symmetric) matrix with $B_i(t)$, $i \in P = \{1, \ldots, p\}$, continuous on I .

Proposition 6.6.1: If f_i, $i = 1, \ldots, p$, is $V-$ invex with respect to α_i, η, $1, \ldots, p$ with $\eta(x, u) = x - u + y(x, u)$, where $B_i y(x, u) = 0$, then $f_i +\cdot^T B_i w$ is also $V-$ invex with respect to η .

Proof: Proof follows easily from the proof of Proposition 2 of Mond and Smart (1988).

In view of Proposition 6.6.1, the Mond-Weir type dual for (NVCP_τ) is the following:

(NVCD_τ) Maximize $\int_a^b \sum_{i=1}^p \tau_i \left(f_i(u) + u^T B_i z_i\right)dt$

subject to $u(a) = \alpha, u(b) = \beta$ (6.38)

$$\sum_{i=1}^p \tau_i \left\{f_x^i(t,u,\dot{u}) + B_i(t)z_i(t)\right\} + \sum_{j=1}^m \lambda_j g_x^j(t,u,\dot{u}) \quad (6.39)$$

$$= \frac{d}{dt}\left\{\sum_{i=1}^p \tau_i \left\{f_{\dot{x}}^i(t,u,\dot{u}) + B_i(t)z_i(t)\right\} + \sum_{j=1}^m \lambda_j g_{\dot{x}}^j(t,u,\dot{u})\right\}$$

$$z_i^T B_i z_i \leq 1, \quad i = 1, \ldots, p \quad (6.40)$$

$$\int_a^b \lambda_j(t)g_j(t,u,\dot{u})dt \geq 0, \quad j = 1, \ldots, m \quad (6.41)$$

$$\lambda(t) \geq 0, \ t \in I, \quad \tau e = 1, \ \tau \geq 0, \quad (6.42)$$

Now Theorem 6.2.1 and Theorem 6.2.2 motivate us to define the following vector maximization variational problem:

(NVCD)

Maximize $\left(\int_a^b \{f_1(t,u,\dot{u}) + u^T B_1 z_1\} dt, ..., \int_a^b \{f_p(t,u,\dot{u}) + u^T B_p z_p\} dt \right)$

subject to (6.38)-(6.42)

Let K and H denote the sets of feasible solutions of (NVCP) and (NVCD), respectively.

Theorem 6.6.3 (Weak Duality):

Let $x \in K$ and $\left(u, \tau, \lambda, z_1, ..., z_p \right) \in H$ .If

$$\left(\int_a^b \{\tau_1 f_1(t,\cdot,\cdot,\cdot) + {}^T B_1(t) z_1(t)\} dt, ..., \int_a^b \{\tau_p f_p(t,\cdot,\cdot,\cdot) + {}^T B_p(t) z_p(t)\} dt \right)$$

are V − pseudo-invex and $\left(\int_a^b \lambda_1 g_1(t,\cdot,\cdot,\cdot) dt, ..., \int_a^b \tau_m g_m(t,\cdot,\cdot,\cdot) dt \right)$ are

V − quasi-invex. Then the following can not hold:

$$\int_a^b \left\{ f_i(t,x,\dot{x}) + \left(x^T B_i x \right)^{\frac{1}{2}} \right\} dt \le \int_a^b \{f_i(t,u,\dot{u}) + u^T B_i z_i\} dt, \quad \forall \; i = 1,...,p$$

$$\int_a^b \left\{ f_{i_0}(t,x,\dot{x}) + (x^T B_{i_0} x)^{\frac{1}{2}} \right\} dt \le \int_a^b \{f_{i_0}(t,u,\dot{u}) + u^T B_{i_0} z_{i_0}\} dt$$

for some $i_0 \in \{1,...,p\}$

Proof: By the feasibility and since $\beta_j(t,x,u) > 0, \quad \forall \; j = 1,...,m,,$
we get

$$\int_a^b \sum_{j=1}^m \beta_j(t,x,u) \lambda_j g_j(t,x,\dot{x}) dt \le \int_a^b \sum_{j=1}^m \beta_j(t,x,u) \lambda_j g_j(t,u,\dot{u}) dt.$$

Then by V − quasi-invexity of

$$\left(\int_a^b \lambda_1 g_1(t,\cdot,\cdot,\cdot) dt, ..., \int_a^b \tau_m g_m(t,\cdot,\cdot,\cdot) dt \right),$$

we get

$$\int_a^b \sum_{j=1}^m \lambda_j \left\{ \eta(t,x,u) g_x^j(t,u,\dot{u}) + \frac{d}{dt} \eta(t,x,u) g_u^j(t,u,\dot{u}) \right\} dt \le 0. \qquad (6.43)$$

From (6.39), we have

$$\int_a^b \eta(t,x,u)\left[\sum_{i=1}^p \tau_i f_x^i(t,u,\dot{u}) + B_i z_i + \sum_{j=1}^m \lambda_j g_x^j(t,u,\dot{u})\right]dt$$

$$= \int_a^b \eta(t,x,u)\frac{d}{dt}\left[\sum_{i=1}^p \tau_i f_{\dot{x}}^i(t,u,\dot{u}) + \sum_{j=1}^m \lambda_j g_{\dot{x}}^j(t,u,\dot{u})\right]dt$$

$$= \eta(t,x,u)\left[\sum_{i=1}^p \tau_i f_{\dot{x}}^i(t,u,\dot{u}) + B_i z_i + \sum_{j=1}^m \lambda_j g_{\dot{x}}^j(t,u,\dot{u})\right]$$

$$- \int_a^b \frac{d}{dt}\eta(t,x,u)\left[\sum_{i=1}^p \tau_i f_{\dot{x}}^i(t,u,\dot{u}) + \sum_{j=1}^m \lambda_j g_{\dot{x}}^j(t,u,\dot{u})\right]dt$$

(by integration by parts).
Thus

$$\int_a^b \eta(t,x,u)\left[\sum_{i=1}^p \tau_i f_x^i(t,u,\dot{u}) + B_i z_i + \sum_{j=1}^m \lambda_j g_x^j(t,u,\dot{u})\right]dt \qquad (6.44)$$

$$+ \int_a^b \eta(t,x,u)\frac{d}{dt}\left[\sum_{i=1}^p \tau_i f_{\dot{x}}^i(t,u,\dot{u}) + \sum_{j=1}^m \lambda_j g_{\dot{x}}^j(t,u,\dot{u})\right]dt$$

$$+ \int_a^b \frac{d}{dt}\eta(t,x,u)\left[\sum_{i=1}^p \tau_i f_{\dot{x}}^i(t,u,\dot{u}) + \sum_{j=1}^p \lambda_j g_{\dot{x}}^i(t,u,\dot{u})\right]dt = 0$$

Since $\eta(t,u,\dot{u}) = 0$ from (6.44), we have

$$\int_a^b \sum_{j=1}^m \left\{\eta(t,x,u)\lambda_j g_x^j(t,u,\dot{u}) + \frac{d}{dt}\eta(t,x,u)\lambda_j g_{\dot{x}}^j(t,u,\dot{u})\right\}dt \qquad (6.45)$$

$$= -\int_a^b \sum_{i=1}^p \left\{\eta(t,x,u)\tau_i f_x^i(t,u,\dot{u})\right.$$

$$\left. + B_i z_i + \frac{d}{dt}\eta(t,x,u)\tau_i f_{\dot{x}}^i(t,u,\dot{u})\right\}dt$$

From (6.45) and (6.43), we have

$$\int_a^b \sum_{i=1}^p \left\{\eta(t,x,u)\tau_i f_x^i(t,u,\dot{u})\right. \qquad (6.46)$$

$$\left. + B_i z_i + \frac{d}{dt}\eta(t,x,u)\tau_i f_{\dot{x}}^i(t,u,\dot{u})\right\}dt \geq 0$$

By V − pseudo-invexity of

$$\left(\int_a^b \left\{\tau_1 f_1(t,\cdot,\cdot,\cdot)+\cdot^T B_1(t)z_1(t)\right\}dt,....,\int_a^b \left\{\tau_p f_p(t,\cdot,\cdot,\cdot)+\cdot^T B_p(t)z_p(t)\right\}dt\right)$$

we have

$$\int_a^b \sum_{i=1}^p \tau_i \alpha_i(x,u,\dot{x},\dot{u})\left\{f_i(t,x,\dot{x})+\left(x^T(t)B_i(t)z_i(t)\right)\right\}dt$$

$$\geq \int_a^b \sum_{i=1}^p \tau_i \alpha_i(x,u,\dot{x},\dot{u})\left\{f_i(t,x,\dot{x})+\left(u^T(t)B_i(t)z_i(t)\right)\right\}dt.$$

By using generalized Schwarz inequality, we get

$$\int_a^b \sum_{i=1}^p \tau_i \alpha_i(x,u,\dot{x},\dot{u})\left\{f_i(t,x,\dot{x})+\left(x^T(t)B_i(t)x(t)\right)^{\frac{1}{2}}\right\}dt$$

$$\geq \int_a^b \sum_{i=1}^p \tau_i \alpha_i(x,u,\dot{x},\dot{u})\left\{f_i(t,x,\dot{x})+\left(u^T(t)B_i(t)u(t)\right)^{\frac{1}{2}}\right\}dt.$$

The conclusion now follows, since $\tau e = 1$ and $B_i(t,x,u) > 0$.

Proposition 6.6.2: Let $u \in K$ and $\left(u,\tau,\lambda, z_1, ..., z_p\right) \in H$. Let the $V-$pseudo-invexity and $V-$quasi-invexity conditions of Theorem 6.2.3 hold. If

$$\left(u^T B_i u\right)^{\frac{1}{2}} = uB_i z, \quad \forall \ i=1,...,p, \tag{6.47}$$

Then u is conditionally properly efficient for (NVCP) and $\left(u,\tau,\lambda, z_1, ..., z_p\right)$ is conditionally properly efficient for (NVCD).

Proof: From (6.46) and (6.47) it follows that for all $x \in K$

$$\int_a^b \sum_{i=1}^p \tau_i \left\{f_i(t,u,\dot{u})+\left(u^T(t)B_i(t)u(t)\right)^{\frac{1}{2}}\right\}dt \tag{6.48}$$

$$= \int_a^b \sum_{i=1}^p \tau_i \left\{f_i(t,u,\dot{u})+\left(u^T(t)B_i(t)z_i(t)\right)\right\}dt.$$

$$\leq \int_a^b \sum_{i=1}^p \tau_i \left\{f_i(t,x,\dot{x})+\left(x^T(t)B_i(t)x(t)\right)^{\frac{1}{2}}\right\}dt \ .$$

Thus u is an optimal solution for the scalarized problem (NCVP$_\tau$). Hence by Theorem 6.6.1, u is a conditionally properly efficient solution for (NVCP).

We first show that $(u, \tau, \lambda, z_1, ..., z_p)$ is an efficient solution for (NVCD). Assume that it is not efficient, i.e., there exists $(\overline{u}, \overline{\tau}, \overline{\lambda}, \overline{z}_1, ..., \overline{z}_p) \in H$ such that

$$\int_a^b \left\{ f_i(t, \overline{u}, \dot{\overline{u}}) + \left(\overline{u}^T(t) B_i(t) \overline{z}_i(t) \right) \right\} dt$$

$$\geq \int_a^b \left\{ f_i(t, u, \dot{u}) + \left(u^T(t) B_i(t) z_i(t) \right) \right\} dt, \forall i = 1, ..., p$$

and

$$\int_a^b \left\{ f_j(t, \overline{u}, \dot{\overline{u}}) + \left(\overline{u}^T(t) B_j(t) \overline{z}_j(t) \right) \right\} dt$$

$$> \int_a^b \left\{ f_j(t, u, \dot{u}) + \left(u^T(t) B_j(t) z_j(t) \right) \right\} dt, \text{ for some } j \in \{1, ..., p\}$$

Thus, from (6.47), we get

$$\int_a^b \left\{ f_i(t, u, \dot{u}) + \left(u^T(t) B_i(t) u(t) \right)^{\frac{1}{2}} \right\} dt$$

$$\leq \int_a^b \left\{ f_i(t, \overline{u}, \dot{\overline{u}}) + \left(\overline{u}^T(t) B_i(t) \overline{z}_i(t) \right) \right\} dt, \forall i = 1, ..., p$$

and

$$\int_a^b \left\{ f_j(t, u, \dot{u}) + \left(u^T(t) B_j(t) u(t) \right)^{\frac{1}{2}} \right\} dt$$

$$< \int_a^b \left\{ f_j(t, \overline{u}, \dot{\overline{u}}) + \left(\overline{u}^T(t) B_j(t) \overline{z}_j(t) \right) \right\} dt, \text{ for some } j \in \{1, ..., p\}$$

contradicting weak duality. Hence $(u, \tau, \lambda, z_1, ..., z_p)$ is efficient.

Now we show that $(u, \tau, \lambda, z_1, ..., z_p)$ is conditionally properly efficient for (NVCD). Assume that it is not conditionally properly efficient i.e., there exist $(\overline{u}, \overline{\tau}, \overline{\lambda}, \overline{z}_1, ..., \overline{z}_p) \in H$ such that for some i and all $M(\overline{u}) > 0$,

$$\int_a^b \left\{ f_i(t,\overline{u},\dot{\overline{u}}) + \left(\overline{u}^T(t)B_i(t)\overline{z}_i(t) \right) \right\} dt \tag{6.49}$$

$$- \int_a^b \left\{ f_i(t,u,\dot{u}) + \left(u^T(t)B_i(t)z_i(t) \right) \right\} dt$$

$$> \int_a^b M(\overline{u}) \left\{ f_j(t,u,\dot{u}) + \left(u^T(t)B_j(t)z_j(t) \right) \right\} dt$$

$$- \int_a^b M(\overline{u}) \left\{ f_j(t,\overline{u},\dot{\overline{u}}) + \left(\overline{u}^T(t)B_j(t)\overline{z}_j(t) \right) \right\} dt$$

for all $j \in \{1,...,p\}$ such that

$$\int_a^b \left\{ f_j(t,u,\dot{u}) + \left(u^T(t)B_j(t)z_j(t) \right) \right\} dt > \int_a^b \left\{ f_j(t,\overline{u},\dot{\overline{u}}) + \left(\overline{u}^T(t)B_j(t)\overline{z}_j(t) \right) \right\} dt \ .$$

Since $\overline{\tau} \geq 0$, $\overline{\tau} \neq 0$,

$$\int_a^b \sum_{i=1}^p \overline{\tau}_i \left\{ f_i(t,\overline{u},\dot{\overline{u}}) + \left(\overline{u}^T(t)B_i(t)\overline{z}_i(t) \right) \right\} dt \tag{6.50}$$

$$> \int_a^b \sum_{i=1}^p \overline{\tau}_i \left\{ f_i(t,u,\dot{u}) + \left(u^T(t)B_i(t)z_i(t) \right) \right\} dt.$$

Now from (6.50) and (6.47), we get

$$\int_a^b \sum_{i=1}^p \overline{\tau}_i \left\{ f_i(t,\overline{u},\dot{\overline{u}}) + \left(\overline{u}^T(t)B_i(t)\overline{z}_i(t) \right) \right\} dt$$

$$> \int_a^b \sum_{i=1}^p \overline{\tau}_i \left\{ f_i(t,u,\dot{u}) + \left(u^T(t)B_i(t)u(t) \right)^{\frac{1}{2}} \right\} dt.$$

contradicting (6.46). Thus $\left(u,\tau,\lambda,z_1,....,z_p \right)$ is conditionally properly efficient.

Theorem 6.6.4 (Strong Duality): Let the $V-$ pseudo-invexity and $V-$ quasi-invexity conditions of Theorem 6.2.3 hold. Let x^0 be normal and a conditionally properly efficient solution for (NVCP). Then for some $\overline{\tau} \in \Lambda^+$, there exists a piecewise smooth $\lambda^0 : I \rightarrow R^m$ such that $\left(u^0 = x^0, \overline{\tau}, \lambda^0 \right)$ is conditionally properly efficient solution for (NVCD) and

$$\int_a^b \left\{ f_i(t,x^0,\dot{x}^0) + \left(x^{0^T}(t)B_i(t)x^0(t) \right)^{\frac{1}{2}} \right\} dt$$

$$= \int_a^b \left\{ f_i(t,u^0,\dot{u}^0) + \left(u^{0^T}(t)B_i(t)z_i^0(t) \right) \right\} dt, \forall i.$$

Proof: Since x^0 is conditionally properly efficient solution for (NVCP) and generalized $V-$ invexity conditions are satisfied, by Theorem 6.6.2, x^0 is optimal for the scalarized primal problem. Therefore, by Theorem 6.6.3, there exists a piecewise smooth $\lambda^0 : I \rightarrow R^m$ such that for $t \in I$,

$$\sum_{i=1}^p \overline{\tau}_i \left\{ f_x^i \left(t,u^0,\dot{u}^0 \right) + B_i(t)z_i^0(t) \right\} + \sum_{j=1}^m \lambda_j^0 g_x^j \left(t,u^0,\dot{u}^0 \right) \qquad (6.51)$$

$$= \frac{d}{dt} \left\{ \sum_{i=1}^p \overline{\tau}_i f_{\dot{x}}^i \left(t,u^0,\dot{u}^0 \right) + \sum_{j=1}^m \lambda_j^0 g_{\dot{x}}^j \left(t,u^0,\dot{u}^0 \right) \right\}$$

$$\left(x^{0^T} B_i x^0 \right)^{\frac{1}{2}} = x^{0^T} B_i z_i^0, \quad i=1,...,p \qquad (6.52)$$

$$z_i^T B_i z_i^0 \leq 1, \quad i=1,...,p \qquad (6.53)$$

$$\lambda^{0^T}(t) g \left(t,u^0,\dot{u}^0 \right) dt = 0, \qquad (6.54)$$

$$\lambda(t) \geq 0, \ t \in I, \quad \tau e = 1, \quad \tau \geq 0. \qquad (6.55)$$

From (6.51) and (6.55) it follows that $\left(x^0, \overline{\tau}, \lambda^0 \right) \in H$. In view of (6.52), by Proposition 6.6.1, $\left(u^0 = x^0, \overline{\tau}, \lambda^0, z_1^0, ..., z_p^0 \right)$ is a conditionally properly efficient solution for (NVCD).Using (6.52), we have

$$\int_a^b \left\{ f_i(t,x^0,\dot{x}^0) + \left(x^{0^T}(t)B_i(t)x^0(t) \right)^{\frac{1}{2}} \right\} dt$$

$$= \int_a^b \left\{ f_i(t,u^0,\dot{u}^0) + \left(u^{0^T}(t)B_i(t)z_i^0(t) \right) \right\} dt, \forall i.$$

References

Aghezzaf B, M Hachimi (2001) Sufficient optimality conditions and duality in multiobjective optimization involving generalized convexity. J Numerical Functional Analysis and Optimization 22:775-788.

Avriel M (1976) Nonlinear Programming: Theory and Methods, Prentice-Hall, New Jersey.

Avriel M, WE Diewert, S Schaible, I Zang (1988) Generalized Concavity, Mathematical Concepts and Methods in Science and Engineering Vol. 36, Plenum Press, New York.

Bazaraa MS, JJ Goode (1973) On symmetric duality in nonlinear programming, J Operations Research 21:1-9.

Bector CR, Bector MK (1987) On various duality theorems in nonlinear programming. J Journal of Optimization Theory and Applications 53:509-515.

Bector CR, Bector MK, Klassen JE (1977) Duality for a nonlinear programming problem. J Utilitas Math. 11:87-99.

Bector CR, Chandra S, Bector MK (1989) Generalized fractional programming duality: a parametric approach. J Journal of Optimization Theory and Applications 60:243-

Bector CR, S Chandra, V Kumar (1994) Duality for minimax programming involving V-invex functions. J Optimization 30:93-103.

Bector CR, Chandra S, Husain I (1994) Optimality conditions and duality in subdifferentiable multiobjective fractional programming. J Journal of Optimization Theory and Applications 79:105-126.

Bector CR, Chandra S, Durga Prasad MV (1988) Duality in pseudolinear multiobjective programming. J Asia-Pacific J. Oper. Res. 5:150-159.

Bector CR, I Husain (1992) Duality for multiobjective variational problems. J Journal of Mathematical Analysis and Applications 166:214-229.

Ben-Israel B, B Mond (1986) What is invexity. J Journal of Australian Mathematical Society 28 B:1-9.

Ben-Tal A, Zowe J. (1982) Necessary and sufficient optimality conditions for a class of nonsmooth minimization problems. J Math. Programming 24:70-91.

Bhatia D, P Kumar (1995) Multiobjective control problems with generalized invexity. J Journal of Mathematical Analysis and Applications 189:676-692.

Bitran G (1981) Duality in nonlinear multiple criteria optimization problems. J Journal of Optimization Theory and Applications 35:367-406.

Borwein JM (1979) Fractional programming without differentiability. J Math. Programming 11:283-290.

Burke JV (1987) Second order necessary and sufficient conditions for composite nondifferentiable optimization. J Math. Programming 38:287-302.

Cambini A, Castagnoli E, Martein L, Mazzoleni P, Schaible S (1990) Generalized convexity and fractional programming with economic applications. In: Proceedings of the International Workshop held at the University of Pisa, Italy 1988.

Chandra S, Craven BD, Husain I (1985) A class of nondifferentiable continuous programming problems. J Journal of Mathematical Analysis and Applications 107:122-131.

Chandra S, Craven BD, Mond B (1986) Generalized fractional programming duality: a ratio game approach, J Journal of Australian Mathematical Society Ser. B 28:170-180.

Chandra S, Durga Prasad MV (1992) Constrained vector valued games and multiobjective programming. J Opsearch 29:1-10.

Chandra S, Durga Prasad MV (1993) Symmetric duality in multiobjective programming. J Journal of Australian Mathematical Society Ser. B 35:198-206.

Chandra S, Kumar V (1993) Equivalent Lagrangian for generalized fractional programming. J Opsearch 30:193-203.

Chandra S, V Kumar (1995) Duality in fractional minimax programming, J. Journal of Australian Mathematical Society Ser. A 58:376-386.

Chandra S, Mond B, Durga Prasad MV (1988) Constrained ratio games and generalized fractional programming. J Zeitschrift fur Oper. Res. 32:307-314.

Chandra S, Mond B, Durga Smart I (1990) Constrained games and symmetric duality with pseudo-invexity. J Opsearch 27:14-30.

Chankong V, YY Haimes (1983) Multiobjective Decision Making: Theory and Methodology. North-Holland, New York.

Charnes A (1953) Constrained games and linear programming. J Proc. Nat. Acad. Sci. (USA) 30:639-641.

Chen XH (1996) Duality for multiobjective variational problems with invexity J Journal of Mathematical Analysis and Applications 203:236-253.

Chew KL, EU Choo (1984) Pseudolinearity and efficiency. J Mathematical Programming 28:226-239.

Clarke FH (1983) Optimization and Nonsmooth Analysis. Wiley, New York.

Coladas L, Z Li, S Wang (1994) Optimality conditions for multiobjective and nonsmooth minimization in abstract spaces. Bulletin of the Australian Mathematical Society 50:205 – 218.

Corley HW (1985) Games with vector pay-offs. J Journal of Optimization Theory and Applications 47:491-498.

Corley HW (1987) Existence and Lagrange duality for maximization of set valued functions. J Journal of Optimization Theory and Applications 54:489-501.

Courant R, Hilbert D (1948) Methods of Mathematical Physics, Vol. 1. Wiley-Interscience, New York.

Cottle RW (1963) Symmetric dual quadratic programs. J Quart. Appl. Math. 21:237-243.

Craven BD (1978) Mathematical Programming and Control Theory. Chapman and Hall, London.

Craven BD (1981) Invex functions and constrained local minima. Bulletin of the Australian Mathematical Society 24:357-366.

Craven BD (1988) Fractional Programming. Sigma Series in Applied Mathematics, Heldermann Verlag Berlin.

Craven BD (1989) Nonsmooth multiobjective programming. J Numerical Functional Analysis and Optimization 10:49 – 64.

Craven BD (1990) Quasimin and quasisaddle points for vector optimization. J Numerical Functional Analysis and Optimization 11:45-54.

Craven BD (1993) On continuous programming with generalized convexity. J Asia-Pacific J. Oper. Res. 10:219-232.

Craven BD (1995) Control and Optimization. Chapman and Hall, New York.

Craven BD, Glover BM (1985) Invex functions and duality. J J. Austral. Math. Soc. 24:1-20.

Craven BD, Mond B (1976) Sufficient Fritz John optimality conditions for nondifferentiable convex programming. J J. Austral. Math. Soc. Ser. B 19:462-468.

Crouzeix JP (1981) A duality framework in quasi-convexprogramming, Generalized Concavity in Optimization and Economics, Edited by S. Schaible and WT Ziemba, Academic Press, New York, 207-225.

Crouzeix JP, Ferland JA, Schaible S (1983) Duality in generalized fractional programming. J Math. Programming 27:342-354.

Crouzeix JP, Ferland JA, Schaible S (1985) An algorithm for generalized fractional programming. J Journal of Optimization Theory and Applications 47:35-49.

Dafermos S (1990) Exchange price equilibrium and variational inequalities. J Mathematical Programming 46:391-402.

Dantzig GB, Eisenberg E, Cottle RW (1965) Symmetric dual nonlinear programs. J Pacific J. of Math. 15:809-812.

De Finetti B (1949) Sulle stratification converse. J Ann. Mat Pura ed Applicata 30:173-183.

Diewert WE, M Avriel, I Zang (1981) Nine kinds of quasi-concavity and concavity. J J. Economic Theory 25:397-420.

Dinkelbach W (1967). On nonlinear fractional programming. J Management Science 13:492-498.

Dorn WS (1960) A symmetric dual theorem for quadratic programs. J J. Oper. Res. Soc. of Japan 2:93-97.

Egudo RR (1987) Proper efficiency and multiobjective duality in nonlinear programming. J J. Information Optim. Sciences 8:155-166.

Egudo RR (1988) Multiobjective fractional duality. Bull. Austral. Math Soc. 37:367-378.

Egudo RR (1989) Efficiency and generalized convex duality for multiobjective programs. J Journal of Mathematical Analysis and Applications 138:84-94.

Egudo RR, MA Hanson (1987) Multi-objective duality with invexity. J Journal of Mathematical Analysis and Applications 126:469-477.

Egudo RR, Hanson MA (1993) On sufficiency of Kuhn-Tucker conditions in nonsmooth multiobjective programming. FSU Technical Report No. M-888.

Elster KH, R Nehse (1980) Optimality Conditions for Some Nonconvex Problems. Springer-Verlag, New York.

Ewing GM (1977) Sufficient conditions for global minima of suitable convex functionals from variational and control theory. SIAM Review 19/2, 202-220.

Fletcher R (1982) A model algorithm for composite nondifferentiable optimization problems. J Math. Programming 17:67-76.

Fletcher R (1987) Practical Methods of Optimization. Wiley, New York.

Friedrichs KD (1929) Verfahren der variations rechnung des minimum eines integral als das maximum eines anderen ausdruckes daeziestellen. Gottingen, Nachrichten.

Geoffrion AM (1968) Proper efficienency and the theory of vector maximization. J Journal of Mathematical Analysis and Applications 22:618-630.

Giorgi G, A Guerraggio, J Thierfelder (2004) Mathematics of Optimization: Smooth and Nonsmooth Case. Elsevier Science B. V., Amsterdam.

Gulati TR, Islam MA (1994) Sufficiency and duality in multiobjective programming involving generalized F-convex functions. J Journal of Mathematical Analysis and Applications 186:181-195.

Gulati TR, Talat N (1991) Sufficiency and duality in nondifferentiable multiobjective programming. J Opsearch 28:73-87.

Hachimi M, B Aghezzaf (2004) Sufficiency and duality in differentiable multiobjective programming involving generalized type I functions. J Journal of Mathematical Analysis and Applications 296:382-392.

Hanson MA (1961) A duality theorem in nonlinear programming with nonlinear constraints. J Austral. J. Statistics 3:64-71.

Hanson MA (1964) Bounds for functionally convex optimal control problems. J Journal of Mathematical Analysis and Applications 8:84-89.

Hanson MA (1981) On sufficiency of the Kuhn-Tucker conditions. J Journal of Mathematical Analysis and Applications 80:545-550.

Hanson MA, B Mond (1982) Further generalization of convexity in mathematical programming. J Journal of Information and Optimization Science 3:25-32.

Hanson MA, B Mond (1987a) Necessary and sufficient conditions in constrained optimization. J Mathematical Programming 37:51-58.

Hanson MA, B Mond (1987b) Convex transformable programming problems and invexity. J Journal of Information and Optimization Sciences 8:201-207.

Hanson MA, R Pini, C Singh (2001) Multiobjective programming under generalized type I invexity. J Journal of Mathematical Analysis and Applications 261:562-577.

Hartley R (1985) Vector and parametric programming. J J. Oper. Res. Soc. 36:423-432.

Henig MI (1982) Proper efficiency with respect to cones. J Journal of Optimization Theory and Applications 36:387-407.

Isbell JR, Marlow WH (1965) Attribtion games. J Nav. Res. Log. Quart. 3:71-93.

Ioffe AD (1979) Necessary and sufficient conditions for a local minimum 2: Conditions of Levin-Milutin-Osmoloviskii type. J SIAM J. Contr. Optim. 17:251-265.

Islam SMN, BD Craven (2005) Some extensions of nonconvex economic modeling: invexity, quasimax and new stability conditions. J Journal of Optimization Theory and Applications 125:315-330.

Ivanov EH, R Nehse (1985) Some results on dual vector optimization problems. J Optimization 16:505-517.

Jahn J (1984) Scalarization in multiobjective optimization. J Math. Programming 29:203-219.

Jahn J (1994) Introduction to the theory of nonlinear optimization. Springer-Verlag Berlin Heidelberg.

Jagannathan R (1973) Duality for nonlinear fractional programs. J Z. Operations Res. Ser. A-B 17 (1973), no. 1:A1--A3.

Jensen JLW (1906) Sur les functionsconvexes et les inegalites entre les valeurs moyennes. J Acta Mathematica 301:75-193.

Jeyakumar V (1987) On optimality conditions in nonsmooth inequality constrained minimization. J Numer. Funct. Anal. Optim. 9:535-546.

Jeyakumar V (1991) Composite nonsmooth programming with Gateauz differentiability. J SIAM J. Optim. 1:30-41.

Jeyakumar V, B Mond (1992) On generalized convex mathematical programming. J Journal of Australian Mathematical Society Ser. B 34:43-53.

Jeyakumar V, Yang XQ (1993) Convex composite multiobjective nonsmooth programming. J Mathematical Programming 59:325-343.

Jeyakumar V, XQ Yang (1995) On characterizing the solution sets of pseudolinear programs. J Journal of Optimization Theory and Applications 87:747-755.

John F (1948) Extremum problems with inequalities as subsidiary conditions. Studies and Essays, Inter-science, New York, 187-204.

Karamardian S (1967) Duality in mathematical programming. J Journal of Mathematical Analysis and Applications 20:344-358.

Kanniappan P (1983) Necessary-conditions for optimality of nondifferentiable convex multiobjective programming. J Journal of Optimization Theory and Applications 40:167-174.

Karlin S (1959) Mathematical methods and theory in games programming and economics, Vol. I, II, Addison-Wesley, Reading Mass.

Karush W (1939) Minima of Functions of several Variables with Inequalities as Side Conditions, M. Sc. Thesis, Department of Mathematics, University of Chicago.

Kaul RN, S Kaur (1985) Optimality criteria in nonlinear programming involving nonconvex functions. J Journal of Mathematical Analysis and Applications 105:104-112.

Kaul RN, Lyall V (1989) Anote on nonlinear fractional vector maximization. J Opsearch 26:108-121.

Kaul RN, SK Suneja, CS Lalitha (1993) Duality in pseudolinear multiobjective fractional programming. J Indian Journal of Pure and Applied Mathematics 24:279-290.

Kaul RN, SK Suneja, MK Srivastava (1994) Optimality criteria and duality in multiple objective optimization involving generalized invexity. J Journal of Optimization Theory and Applications 80:465-482.

Kawaguchi T, Maruyama Y (1976) A note on minimax (maximin) programming. J Management Sci. 22:670-676.

Kreindler E (1966) Reciprocal optimal control problems. J Journal of Mathematical Analysis and Applications 14:141-152.

Kim DS, AL Kim (2002) Optimality and duality for nondifferentiable multiobjective variational problems. J Journal of Mathematical Analysis and Applications 274:255-278.

Kim DS, WJ Lee (1998) Symmetric duality for multiobjective variational problems with invexity. J Journal of Mathematical Analysis and Applications 218:34-48.

Kim MH, GM Lee (2001) On duality theorems for nonsmooth Lipschitz optimization problems. J Journal of Optimization Theory and Applications 110:669-675.

Kim DS, GM Lee, JY Park, KH Son (1993) Control problems with generalized invexity. J Math. Japon. 38:263-269.

Kim DS, WJ Lee, S Schaible (2004) Symmetric duality for invex multiobjective fractional variational problems. J Journal of Mathematical Analysis and Applications 289:505-521.

Kim DS, YB Yun, WJ Lee (1998) Multi-objective symmetric duality with cone constraints. J European Journal of Operational Research 107:686-691.

Komlosi S (1993) First and second order characterizations of pseudolinear functions. J European Journal of Operational Research 67:278-286.

Komlosi S, Rapesak T, Schaible S(eds.) (1994) Generalized convexity. In: Proceedings Pecs/Hungary, 1992; Lecture Notes in Economics and Mathematical Systems 405, Springer-Verlag,berlin-Heidelberg, New York.

Kuhn HW (1976) Nonlinear programming: a historical view. In: R.W. Cottle, C. E. Lemke (eds.) Nonlinear Programming, SIAM-AMS Proceedings 9:1-26.

Kuhn HW, AW Tucker (1951) Nonlinear programming. In: J. Neyman (ed.): Proceedings of the Second Berkeley Symposium on Mathematical Statistics and Probability, University of California Press, Berkeley, 481-492.

Lal SN, Nath B, Kumar A (1994) Duality for some nondifferentiable static multiobjective programming problems. J Journal of Mathematical Analysis and Applications 186:862-867.

Lai HC, Ho CP (1986) Duality theorem of nondifferentiable convex multiobjective programming. J Journal of Optimization Theory and Applications 50:407-420.

Lai HC, JC Lee (2002) On duality theorems for a nondifferentiable minimax fractional programming. J Journal of Computational and Applied Mathematics 146:115-126.

Lee GM (1994) Nonsmooth invexity in multiobjective programming". J Journal of Information and Optimization Sciences 15:127 – 136.

Luo HZ, ZK Xu (2004) On characterization of prequasi-invex functions. J Journal of Optimization Theory and Applications 120:429-439.

Maeda T (1994) Constraint qualifications in multiobjective optimization problems: differentiable case. J Journal of Optimization Theory and Applications 80:483—500.

Mancino OG, G Stampacchia (1972) Convex programming and variational inequalities. J Journal of Optimization Theory and Applications 9:3-23.

Mangasarian OL (1965) Pseudo-convex functions. J SIAM Journal on Control 3:281-290.

Mangasarian OL (1969) Nonlinear Programming. McGraw-Hill, New York.

Martin DH (1985) The essence of invexity. J Journal Optimization Theory and Applications 47:65-76.

Marusciac I (1982) On Fritz John type optimality criterion in multiobjective optimization. J L'Analyse Numerique et la Theorie de l'Approximation 11:109-114.

Mastroeni G (1999) Some remarks on the role of generalized convexity in the theory of variational inequalities. In: G. Giorgi and F. Rossi (eds.) Generalized Convexity and Optimization for Economic and Financial decisions, Pitagora Editrice Bologna, 271-281.

Mazzoleni P(ed.) (1992) Generalized concavity. In: Proceedings, Pisa, April 1992, Technopnut S. N. C., Bologna.

Mishra SK (1995) Pseudolinear fractional minimax programming. J Indian J. Pure Appl. Math. 26:763-772.

Mishra SK (1996a) On sufficiency and duality for generalized quasiconvex nonsmooth programs. J Optimization 38:223 – 235.

Mishra SK (1996b) Generalized proper efficiency and duality for a class of nondifferentiable multiobjective variational problems with V-invexity. J Journal of Mathematical Analysis and Applications 202:53-71.

Mishra SK (1996c) Lagrange multipliers saddle points and scalarizations in composite multiobjective nonsmooth programming. J Optimization 38:93-105.

Mishra SK (1997b) On sufficiency and duality in nonsmooth multiobjective programming. J Opsearch 34:221-231.

Mishra SK (1998a) Generalized pseudoconvex minimax programming. J Opsearch 35:32-44.

Mishra SK (1998b) On multiple objective optimization with generalized univexity. J Journal of Mathematical Analysis and Applications 224:131-148.

Mishra SK (2000a) Multiobjective second order symmetric duality with cone constraints. J European Journal of Operational Research 126:675-682.

Mishra SK, G Giorgi (2000) Optimality and duality with generalized semiunivexity. J Opsearch 37:340-350.

Mishra SK, G Giorgi, SY Wang (2004) Duality in vector optimization in Banach spaces with generalized convexity. J Journal of Global Optimization 29:415-424.

Mishra SK, RN Mukherjee (1994a) Duality for multiobjective fractional variational problems. J Journal of Mathematical Analysis and Applications 186:711-725.

Mishra SK, RN Mukherjee (1994b) On efficiency and duality for multiobjective variational problems. J Journal of Mathematical Analysis and Applications 187:40-54.

Mishra SK, RN Mukherjee (1995a) Generalized continuous nondifferentiable fractional programming problems with invexity. J Journal of Mathematical Analysis and Applications 195:191-213.

Mishra SK, RN Mukherjee (1995b) Generalized convex composite multiobjective nonsmooth programming and conditional proper efficiency. J Optimization 34:53-66.

Mishra SK, RN Mukherjee (1996) On generalized convex multiobjecive nonsmooth programming. J Journal of the Australian Mathematical Society B 38:140 – 148.

Mishra SK, RN Mukherjee (1997) Constrained vector valued ratio games and generalized subdifferentiable multiobjective fractional minmax programming. J Opsearch 34:1-15.

Mishra SK, RN Mukherjee (1999) Multiobjective control problems with V-invexity. J Journal of Mathematical Analysis and Applications 235:1-12.

Mishra SK, MA Noor (2005) On vector variational-like inequality problems. J Journal of Mathematical Analysis and Applications 311:69-75

Mishra SK, NG Rueda (2000) Higher order generalized invexity and duality in mathematical programming. J Journal of Mathematical Analysis and Applications 247:173-182.

Mishra SK, NG Rueda (2002) Higher order generalized invexity and duality in nondifferentiable mathematical programming. J Journal of Mathematical Analysis and Applications 272:496-506.

Mishra SK, NG Rueda (2003) Symmetric duality for mathematical programming in complex spaces with F-convexity. J Journal of Mathematical Analysis and Applications 284:250-265.

Mishra SK, SY Wang (2005) Second order symmetric duality for nonlinear multiobjective mixed integer programming. J European Journal of Operational Research 161:673-682.

Mishra SK, SY Wang, KK Lai (2005) Nondifferentiable multiobjective programming under generalized d-univexity. J European Journal of Operational Research 160:218-226.

Mohan SR, SK Neogy (1995) On invex sets and preinvex functions". J Journal of Mathematical Analysis and Applications 189:901-908.

Mond B (1965) A symmetric dual theorem for nonlinear programs". J Quarterly Journal of Applied Mathematics 23:265-269.

Mond B (1974) A class of non-differentiable mathematical programming problems. J Journal of Mathematical Analysis and Applications 46:169-174.

Mond B, S Chandra, I Husain (1988) Duality for variational problems with invexity. J Journal of Mathematical Analysis and Applications 134:322-328.

Mond B, MA Hanson (1967) Duality for variational problems. J Journal of Mathematical Analysis and Applications 18:355-364.

Mond B, MA Hanson (1968) Duality for control problems. J SIAM J. Control 6:114-120.

Mond B, Hanson MA (1984) On duality with generalized convexity. J Math. Oper. Statist. Ser. Optim. 15:313-317.

Mond B, Husain I, Durga Prasad MV (1991) Duality for a class of nondifferentiable multiobjective programs. J Utilitas Mathematica 39:3-19.

Mond B, I Smart (1988) Duality and sufficiency in control problems with invexity. J Journal of Mathematical Analysis and Applications 136:325-333.

Mond B, I Smart (1989) Duality with invexity for a class of nondifferentiable static and continuous programming problems. J Journal of Mathematical Analysis and Applications 141:373-388.

Mond, B, T Weir (1981) Generalized concavity and duality. In: S Schaible and WT Ziemba, (eds.), Generalized Concavity Optimization and Economics, Academic Press, New York, 263-280.

Mond B, T Weir (1981-1983) Generalized convexity and higher-order duality. J Journal of Mathematical Sciences 16-18:74-94.

Mukherjee RN (1991) Generalized convex duality for multiobjective fractional programs. J Journal of Mathematical Analysis and Applications 162:309-316.

Mukherjee RN, SK Mishra (1994) Sufficiency optimality criteria and duality for multiobjective variational problems with V–invexity. J Indian J. Pure Appl. Math. 25:801 – 813.

Mukherjee RN, SK Mishra (1995) Generalized invexity and duality in multiple objective variational problems. J Journal of Mathematical Analysis and Applications 195:307-322.

Mukherjee RN, SK Mishra (1996) Multiobjective programming with semilocally convex functions. J Journal of Mathematical Analysis and Applications 199:409 – 424.

Nahak C, S Nanda (1996) Duality for multiobjective variational problems with invexity. J Optimization 36:235-258.

Nahak C, S Nanda (1997a) Duality for multiobjective variational problems with pseudoinvexity. J Optimization 41:361-382.

Nahak C, S Nanda (1997b) On efficientand duality for multiobjective variational control problems with (F,ρ)-convexity. J Journal of Mathematical Analysis and Applications 209:415-434.

Nakyam H (1984) Geometric consideration of duality in vector optimization. J Journal of Optimization Theory and Applications 44:625-655.

Nanda S, LN Das (1996) Pseudo-invexity and duality in nonlinear programming. J European Journal of Operational Research 88:572-577.

Nanda S, Das LN (1994) Pseudo-invexity and symmetric duality in nonlinear programming. J Optimization 28:267-273.

Osuna R, A Rufian, G Ruiz (1998) Invex functions and generalized convexity in multiobjective programming. J Journal of Optimization Theory and Applications 98:651-661.

Pearson JD (1965) Reciprocity and duality in control programming problems. Ibid 10:383-408.

Pini R, C Singh (1997) A survey of recent [1985-1995] advances in generalized convexity with applications to duality theory and optimality conditions. J Optimization 39:311-360.

Phuong TD, PH Sach, ND Yen (1995) Strict lower semicontinuity of the level sets and invexity of a locally Lipschitz function. J Journal of Optimization Theory and Applications 87:579 – 594.

Ponstein J (1967) Seven types of convexity. Society for Industrial and Applied Mathematics Review 9:115-119.

Preda V (1992) On efficiency and duality for multi-objective programs. J Journal of Mathematical Analysis and Applications 166:365-377.

Preda V (1994) On sufficiency and duality for generalized quasiconvex programs. J Journal of Mathematical Analysis and Applications 181:77-88.

Rapcsak T (1991) On pseudolinear functions. J European Journal of Operational Research 50:353-360.

Riesz F, B Sz Nagy (1955) Functional Analysis. Frederick Ungar Publishing, New York.

Ringlee RJ (1965) Bounds for convex variational programming problems arising in power system scheduling and control. J IEEE Trans. Automatic Control, Ac-10: 28-35.

Rojas-Medar MA, AJV Brandao (1998) Nonsmooth continuous-time optimization problems: sufficient conditions. J Journal of Mathematical Analysis and Applications 227:305-318.

Rockafellar RT (1969) Convex Analysis. Princeton University Press, Princeton New Jersey.

Rockafellar RT (1988) First and second order epi-differentiability in nonlinear programming. J Trans. Amer. Math. Soc. 307:75-108.

Rodder W (1977) A generalized saddle point theory. J European J. Oper. Res. 1:55-59.

Rosenmuller J, HG Weidner (1974) Extreme convex set functions with finite carries: general theory. J Discrete Math. 10:343-382.

Rueda NG (1989) Generalized convexity in nonlinear programming. J J. Information and Optimization Sciences 10:395-400.

Rueda NG, MA Hanson (1988) Optimality criteria in mathematical programming involving generalized invexity, J Journal of Mathematical Analysis and Applications 130:375-385.

Rueda NG, MA Hanson, C Singh (1995) Optimality and duality with generalized convexity. J Journal of Optimization Theory and Applications 86:491-500.

Ruiz-Garzion G, R Osuna-Gomez, A Ruffian-Lizana (2003) Generalized invex monotonicity. J European Journal of Operational Research 144:501-512.

Ruiz-Garzion G, R Osuna-Gomez, A Rufian-Lizan (2004) Relationships between vector variational-like inequality and optimization problems. J European Journal of Operational Research 157:113-119.

Sawaragi S, Nakayama H, Tanino T (1985) Theory of multiobjective optimization. Academic Press.

Schaible S (1976a) Duality in fractional programming; a unified approach. J Oper. Res. 24:452-461.

Schaible S (1976b) Fractional programming: I, duality. J Management Sci. 22:858-867.

Schaible S (1981) A survey of fractional programming. Generalized Concavity in Optimization and Economics. Edited by S Schaible and WT Ziemba, Academic Press, New York, 417-440.

Schaible S (1995) Fractional programming. In: R Horst, PM Pardalos (Eds.), Handbook of Global Optimization, Kluwer Academic, Dordrecht, 495-608.

Schaible S, WT Ziemba (1981) Generalized Concavity in Optimization and Economics. Academic Press, New York.

Schechter M (1979) More on subgradient duality. J Journal of Mathematical Analysis and Applications 71:251-262.

Schroeder RG (1970) Linear programming solutions to ration games. J Operations Research 18:300-305.

Schmitendorf WE (1977) Necessary conditions and sufficient conditions for static minimax problems. J Journal of Mathematical Analysis and Applications 57:683-693.

Singh C (1986) A class of multiple criteria fractional programming problems. J Journal of Mathematical Analysis and Applications 115:202-213.

Singh C (1988) Duality theory in multiobjective differentiable programming programming. J Journal of Information and Optimization Science 9:231-240.

Singh C, Hanson MA (1986) Saddle point theory for nondifferentiable multiobjective fractional programming. J J. Information and Optimization Sciences 7:41-48.

Singh C, Hanson MA (1991) Generalized proper efficiency in multiobjective fractional programming. J Journal of Information and Optimization Sciences 12:139-144.

Singh C, Rueda NG (1990) Generalized fractional programming and duality theory. J Journal of Optimization Theory and Applications 57:189-196.

Singh C, Rueda NG (1994) Constrained vector valued games and generalized multiobjective minmax programming. J Opsearch 31:144-154.

Smar I (1990) Invex Functions and Their Application to Mathematical Programming. Ph. D. Thesis, La Trobe University, Bundoora, Victoria, Australia.

Smart I, B Mond (1990) Symmetric duality with invexity in variational problems. J Journal of Mathematical Analysis and Applications 152:536-545.

Smart I, B Mond (1991) Complex nonlinear programming: duality with invexity and equivalent real programs. J Journal of Optimization Theory and Applications 69:469-488.

Stancu-Minasian IM (1997) Fractional Programming: Theory, Methods and Applications, Mathematics and Its Application Vol. 409. Kluwer Academic Publishers, Dordrecht.

Stancu-Minasian IM (2002) Optimality and duality in fractional programming involving semilocally preinvex and related functions. J J. Information Optimization Science 23:185-201.

Suneja SK, S Gupta (1998) Duality in multiobjective nonlinear programming involving semilocally convex and related functions. J European Journal of Operational Research 107:675-685.

Suneja SK, CS Lalitha, S Khurana (2003) Second order symmetric duality in multiobjective programming. J European Journal of Operational Research 144:492-500.

Suneja SK, C Singh, CR Bector (1993) Generalization of preinvex and b-vex functions. J Journal of Optimization Theory and Applications 76:577-587.

Suneja SK, Srivastava (1994) Duality in multiobjective fractional programming involving generalized invexity. J Opsearch 31:127-143.

Suneja SK, MK Srivastava (1997) Optimality and duality in nondifferentiable multiobjective optimization involving d-type I and related functions. J Journal of Mathematical Analysis and Applications 206:465-479.

Tamura K, Arai S (1982) On proper and improper efficient solutions of optimal problems with multicriteria. J Journal of Optimization Theory and Applications 38:191-205.

Tanaka T (1988) Some minimax problems of vector valued functions. J Journal of Optimization Theory and Applications 59:505-524.

Tanaka T (1990) A characterization of generalized saddle points for vector valued functions via scalarization. J Nihonkai Math. J. 1:209-227.

Tanaka Y, Fukusima M, Ibaraki I (1989) On generalized pseudo-convex functions. J Journal of Mathematical Analysis and Applications 144:342-355.

Tanimoto S (1981) Duality for a class of nondifferentiable mathematical programming problems. J Journal of Mathematical Analysis and Applications 79:286-294.

Tanino T, Y Sawaragi (1979) Duality theory in multi-objective programming. J Journal of Optimization Theory and Applications 27:509-529.

Tucker AW (1957) Linear and nonlinear programming. J Oper. Res. 5:244-257

Valentine FA (1937) The problem of Lagrange with differential inequalities as added side conditions, Contributions to the Calculus of Variations (1933-1937). University of Chicago Press, Chicago.

Valentine FA (1964) Convex Sets. McGraw-Hill, New York.

Vogel W (1974) Ein maximum prinzip fur vector optimierungs-Aufgaben. J Oper. Res. Verfahren 19:161-175.

Wang SY (1984) Theory of the conjugate duality in multiobjective optimization. J J. System. Sci. and Math. Sci. 4:303-312.

Weir T (1986) A dual for a multiobjective fractional programming problem. J J. Information Optim. Sci. 7:261-269.

Weir T (1991) Symmetric dual multiobjective fractional programming. J Journal of Australian Mathematical Soc. Ser. A 50:67-74.

Weir T (1986a) A duality theorem for a multiple objective fractional optimization problem. J Bull. Austral. Math. Soc. 34:415-425.

Weir T (1987) Proper efficiency and duality for vector valued optimization. J J. Austral. Math. Soc. Ser. A 43:21-34.

Weir T (1988) A note on invex functions and duality in multiple objective optimization. J Opsearch 25:98-104.

Weir T (1989) On duality in multiobjective fractional programming. J Opsearch 26:151-158.

Weir T (1992) Pseudoconvex minimax programming. J Util. Math. 42:234-240.

Weir T, B Mond (1984) Generalized convexity and duality for complex programming problem". J Cahiers de C. E. R. O. 26:137-142.

Weir T, B Mond (1988a). Preinvex functions in multiobjective optimization. J Journal of Mathematical Analysis and applications 136:29-38.

Weir T, B Mond (1988b). Symmetric and self duality in multiple objective programming. J Asia-Pacific Journal of Oper. Res. 5:124-133.

Weir T, B Mond (1989) Generalized convexity and duality in multiple objective programming. J Bulletin of the Australian Mathematical Society 39:287-299.

Weir T, B Mond, Craven BD (1986) On duality for weakly minimized vector valued optimization problems. J Optimization 17:711-721.

Weir T, B Mond, Craven BD (1989) Generalized onvexity and duality in multiple objective programming. J Bull. Austral. Math. Soc. 39:287-299.

White DJ (1985) Vector maximization and Lagrange multipliers. J Math. Programming 31:192-205.

Wolfe, P (1961) A duality theorem for non-linear programming. J Quarterly of Applied mathematics 19:239-244.

Xu ZK (1988) Saddle point type optimality criteria for generalized fractional programming. J Journal of Optimization Theory and Applications 57:189-196.

Xu, Z (1996) Mixed type duality in multiobjective programming problems. J Journal of Mathematical Analysis and Applications 198:621-635.

Yadav SR, RN Mukherjee (1990) Duality for fractional minimax programming problems. J J. Austral. Math. Soc. Ser. B 31:484-492.

Yang XM, KL Teo, XQ Yang (2000) Duality for a class of non-differentiable multi-objective programming problems. J Journal of Mathematical Analysis and Applications 252:999-1005.

Yang XM, KL Teo, XQ Yang (2002a). Symmetric duality for a class of nonlinear fractional programming problems. J Journal of Mathematical Analysis and Applications 271:7-15.

Yang XM, SY Wang, XT Deng (2002b) Symmetric duality for a class of multiobjective fractional programming problems. J Journal of Mathematical Anaysis and Applications 274:279-295.

Yang XM, XQ Yang, KL Teo (2001) Characterization and applications of prequasi-invex functions. J Journal of Optimization Theory and Applications 110:645-668.

Ye YL (1991) D-invexity and optimality conditions. J Journal of Mathematical Analysis and Applications 162:242-249.

Zalmai GJ (1985) Sufficient optimality conditions in continuous-time nonlinear programming. J Journal of Mathematical Analysis and Applications 111:130-147.

Zalmai GJ (1987) Optimality criteria and duality for a class of minimax programming problems with generalized invexity conditions. J Utilitas Math. 32:35-57.

Zalmai GJ (1990b) Generalized sufficient criteria in continuous-time programming with application to a class of variational-type inequalities. J Journal of Mathematical Analysis and Applications 153:331-355.

Zang I, Choo EU, Avriel M (1977) On functions whose stationary points are global minima. J Journal of Optimization Theory and Applications 22:195-208.

Zhao F (1992) On sufficiency of the Kuhn-Tucker condition in nondiferentiable programming. J Bull. Austral. Math. Soc. 46:385-389.

Zhian L (2001) Duality for a class of multiobjective control problems with generalized invexity. J Journal of Mathematical Analysis and Applications 256:446-461.

Subject Index

Author Index